Years of Change

Mike Soper

FARMING PRESS

First published 1995

ISBN 0 85236 313 3

A catalogue record for this book is available
from the British Library

Published by Farming Press Books
Miller Freeman Professional Ltd
Wharfedale Road, Ipswich IP1 4LG, United Kingdom

Distributed in North America
by Diamond Farm Enterprises,
Box 537, Alexandria Bay, NY 13607, USA

Cover design by Smart Graphics
Main cover photograph – Marshall 626 combine harvester –
courtesy of Stuart Gibbard
Photograph inset on front cover, courtesy of Massey Ferguson
Photograph inset on back cover shows Oxford University
farm buildings
Typeset by Galleon Typesetting
Printed and bound in Great Britain by
Biddles Ltd, Guildford and King's Lynn

Contents

v

Preface

———✺———

A detailed historical record of the development of British agriculture over the past 60 years would occupy several volumes. This book is not intended as such a chronicle. I have merely attempted to present a personal account of some of the more significant changes that I have seen and of activities in which I have been involved over this period. Such an autobiographical approach can easily fall into a number of traps. It can include over-much detail about things that the author finds interesting but which are of little real significance to others. It can involve including many names which mean little to the reader, and it can encourage an excessive use of the personal pronoun.

I have tried hard to avoid such pitfalls and would ask for the forbearance of those who may feel that I have failed to do so adequately, and especially of those whose names I would like to have included, but finally omitted for the sake of brevity. I can assure them that they are not forgotten.

M. H. R. Soper

Acknowledgements

———

I would like to express my thanks to the numerous people I consulted when checking facts, and especially to my former colleagues Geoff Hodgson and Ian Morton, and to Professor Tony Giles, who was particularly helpful with economic details. I would also like to take this opportunity to thank the latter for his masterly history of the Oxford Farming Conference, *See You at Oxford*. With the Conference celebrating its fiftieth anniversary in January 1996, the publication of his book in 1994 was very apposite.

Finally, I would like once more to acknowledge the debt that I owe to all the farmers, educationists, scientists and others who, in so many ways, have made this 60-year agricultural pilgrimage such a stimulating, enjoyable and rewarding experience – and are still doing so today.

M. H. R. Soper

CHAPTER ONE

Apprenticeship

LIFE suddenly took on a completely new dimension in July 1934 when I returned to Cambridge for the Long Vacation term to begin the two year postgraduate course for the Diploma in Agriculture. I had just achieved an extremely mediocre degree in natural science by squeezing the normal three year tripos into two years.

Life took on a new dimension because up to that time I had rather meandered through school and two years at university doing subjects which, with the exception of biology at school and physiology in my degree, had not enthused me in any way. I had had up till then, at considerable cost to my parents, to whom I will always be grateful, the conventional middle class education of the time: a boarding school on the South Coast from eight to fourteen, five years at a very good public school in Kent, followed by Cambridge which had still left me with no very clear idea of where I wanted to go, except that I did *not* want a job in an office, I did *not* want a job which involved physics and mathematics, and I wanted to be able to get out of doors as much as possible.

There were three things which influenced my father in pushing me gently in the direction of agriculture, when it became necessary to decide what to read at Cambridge. Yes, he had to do the pushing as I was in reality rather a late developer, and not confident enough to take decisions off my own bat, even at 18. In the first place, my brother, who was nine years older than me, was in the Colonial Agricultural Service, where he was doing well. In addition, my father's small manufacturing firm had been taken over by a subsidiary of ICI, and he was aware through contacts in the company that they were expanding quite rapidly in the agricultural field at the time, so that there might be jobs in research or advisory work available there. Finally, my paternal grandmother had come from a long line of Essex farmers and merchants (the Philpots, still actively farming there today), and there were undoubtedly strong 'farming genes' in the family.

Not actual farming genes, I suppose, but inherited genetic factors capable of inducing in the individual a strong affinity with the natural world, and living organisms and their care. I suspect these had become dominantly fixed over many previous generations of a farming family dating back to the Huguenots in 1685, and probably before that, in France. How else can one account for the fact that, at the age of 20, with no previous background or experience of agriculture whatsoever, I suddenly felt on starting the diploma course at Cambridge completely at home with the environment in which I was working? I was confident, for once in my life, that this was what I was meant to do, and I felt in some strange way that I already knew a great deal about it.

My father had been advised by his ICI contacts that the best preparation for a professional agricultural career was to cover the Cambridge natural science degree course in two years, and then the two year diploma course rather than the ordinary three year degree course in agriculture. This was very good advice, but it did mean doing a fourth year at university, and an additional burden on parental finances which were very strained, as my father had already retired by then, and this was in the days before student grants were available.

Some additional finance did fortunately become available to alleviate the situation for my fourth year.

THE CAMBRIDGE DIPLOMA

I was especially fortunate to have entered the Cambridge School of Agriculture when I did, as it was probably at its zenith in the mid 1930s. University departments, and this is probably true of most organisations of a similar kind, tend to go through phases of excellence and distinction for a number of years and then regress for a period, after which they may bounce back, or possibly go into a permanent decline, which is unfortunately what ultimately happened at Cambridge with agriculture some 30 years later. The main factor which influences this is the personality and ability of the man at the top, or, it may be, several men or women at the top. An outstanding professor will attract the best available talent to work with him either as research workers, or teachers, or both. Their reputation then attracts the best calibre junior researchers, which further enhances the reputation of that school, and if it is a teaching department as well, good ordinary degree students are also attracted to the courses offered, because of the proven quality of the academic staff.

This had happened at Cambridge before the First World War when a number of talented scientists had come together, and were applying science in an enlightened way to the problems of plant and animal production. Among this group were Professor T. B. Wood, who laid the foundations for the scientific rationing of livestock, Sir Rowland Biffen, the cereal breeder responsible for several of the early higher yielding varieties of wheat, the young John Hammond, the animal physiologist, and many others. The School continued to expand in the '20s, when H. G. Sanders came to work with John Hammond as a PhD student, later becoming a lecturer in crop production, W. S. Mansfield arrived as Farm Director, and G. D. H. Bell followed Biffen as an outstanding breeder of cereal crops. Frank Garner joined the staff in the late '20s, and H. E. Woodman's research led to that early Bible of the feeding world, *Rations for Livestock*.

It is perhaps a measure of the calibre and national status of the teaching staff that by the time I arrived as a student in 1934, of the ten people who lectured to me, three were subsequently to receive knighthoods (Engledow, Hammond and Sanders), and two were to receive CBEs (Mansfield and Bell) for their services to the agricultural industry in their respective fields. This must be an all time record for an agricultural institution.

The spin-off from this success was that by the early 1920s agriculture had become recognised in the university as a reputable scientific and academic subject on a par with botany and zoology, rather than something to be read by the landed gentry or the idle rich who were seeking a convenient vehicle for spending three pleasurable but unproductive years at Cambridge. Outside bodies were prepared to provide funds for research, and, in addition, other institutions, such as the National Institute of Agricultural Botany (NIAB) were established in the immediate vicinity. It was a case of success breeding success.

With an increasing amount of research being centred on the School and also being carried out nationally, there was felt to be a need for a postgraduate qualification for those likely to follow careers in research, education or administration. At the same time, the Colonial Office, as it then was, became concerned at the variable scientific backgrounds and qualifications of its recruits for the Colonial Agricultural Service, and felt the need for an initial training course to bring each intake to a more uniform standard in scientific agriculture before they were let loose in a developing country.

So a diploma course was started in 1930, jointly sponsored by the university and the Colonial Office, but the Colonial scholars only attended the first year, moving on for a second year to the College of Tropical Agriculture (in Trinidad).

The great merits of the course were its small size, the highly personal nature of the instruction, and the very strong staff/student relationship. In my first year, there were only three of us coming in from the Cambridge degree course, and, in addition, some ten colonial scholars. Each student was allocated a major experiment on the farm as his personal responsibility, and this constituted the subject of the dissertation which had to be submitted by the end of the second year as part of the final examination. We worked as a team on the fieldwork for each other's experiments, and analysed the results jointly as a class, so that one came to know almost as much about the other trials as one did about one's own. They were spread over both crop and animal topics, and proved an excellent method for acquiring information across quite a wide field.

My own particular trial was on the nitrogenous manuring of permanent pasture, and its effect on the sward constituents, and it is fascinating to look back at the report, and to see the very small quantities of fertiliser which were thought to be appropriate at that time.

The course tutors were Harold Sanders, always known even then as 'Jem' Sanders, though I never found out the origin of this name, and Frank Garner, and a remarkably effective team they had become as the course matured. They had completely different personalities. Sanders with a first class scientific and mathematical brain, very extrovert and highly approachable, and full of dynamic energy was a born natural leader, as his subsequent career was to prove. Frank Garner, on the other hand, with the external appearance of a very solid and dependable Hereford bull, was a much more serious character. Somewhat introverted in the company of Sanders (as many people were) he was a very competent agriculturalist, a sound scientist, and a very kindly and thoughtful man behind his apparent reserve, when one got to know him well. They complemented one another admirably on the course, with Garner playing the role of staff officer to his senior partner most effectively. Sanders was certainly the 'ideas' man, who led from the front, but it was often Garner that one would go to for elaboration of some point that one did not understand, or a statistical calculation in which one was bogged down (as was often the case with me, with my mathematical ineptitude).

The debt that I owe to these two is quite incalculable, as they set an example in personal relationships which I have done my best to follow throughout my subsequent career, and they had a tremendous influence on my approach to agriculture at a most impressionable moment in my intellectual development. When Jem Sanders died in 1985, just before his 87th birthday, I was delighted to be asked, firstly, to take the lead in organising a memorial service for him in

London, and secondly, to write an appreciation of his life and contribution to agriculture (which was published in the *RASE Journal* in 1986) as I felt that this went a little way to repaying him for giving me such an inspiring start to my career in agriculture.

On the diploma course, we attended Sanders' and Garner's lectures to the second and third years of the agriculture degree course, which provided those of us who were pretty green with regard to farming, with good factual material on the practicalities of crop and animal production. They still adhered to the strange arrangement, current at the time, of divorcing crops from animals, and dealing mainly with crops and grassland in the second year of the course, and animals in the third year. I found this again later when I moved to Reading University, but took good care to see that the two subjects were taught concurrently when I went to Oxford after the War, as it seemed to me completely illogical to try to divorce them, when dealing with such closely integrated systems of production as are found in farming. We doubled up with the degree course for a few subjects, but were taught separately in the applied sciences, such as nutrition, animal physiology, genetics and economics, since we were working at a more advanced level. This applied especially to biological statistics which was a very strong feature of the course.

The latter subject was still in its infancy, as it had not really been applied to agriculture till after the First World War, when R. A. Fisher and A. F. Yates began to develop the principles and apply them experimentally at Rothamsted. Sanders, with his mathematical and enquiring mind, immediately recognised the importance of applying standard methods of analysis to the results of agricultural experiments, whether with plants or animals, and he had, in fact, spent a year on secondment working with Fisher at Rothamsted in the early 1920s. So this subject became an integral part of the course, since it was recognised that a comprehensive knowledge of the subject would need to be part of the equipment of research workers in agriculture as research expanded in the future. I must confess, now that it no longer matters, that I found it rather ironic that I had chosen agriculture as a career as far away from mathematics as I could imagine, only to find myself enmeshed in fairly complicated mathematical methods that I found very difficult to comprehend, and which I never really mastered.

However, I managed to get by somehow, because Sanders taught it largely through the analysis of experiments on the farm, and these provided, to my materialistic mind, enough practical reality to make the working out of results reasonably easy provided that they fell within a clearly defined set of rules. But I never did understand the mathematical principles well enough to be able to adapt an analysis for an atypical set of results, which is the main reason, I expect, why I never became a dedicated research worker.

That first eight week term was spent partly in getting to grips with the basics of agriculture in the lecture room and the laboratory and partly with harvesting the previous year's cereal, bean and grassland experiments. The summer of 1934 was a hot one, and I have quite vivid memories of the student harvest gang led as usual from the front by Jem Sanders stripped to the waist, cutting plots with hooks, binding and weighing, and then stooking the sheaves, for subsequent analysis in the laboratory. This working together as a group in the field with our tutors was a very enlightening, and rewarding experience. It was the first time that we had been treated as complete equals by someone in authority, and it was a considerable confidence booster to be regarded by them as mature and responsible individuals. In the previous degree course, the large numbers involved in nearly all the subjects precluded any opportunity of building up any personal contacts with lecturers or tutors.

Towards the end of the summer term in the first year, Jem Sanders asked me to see him one day, and offered me the job of Junior Assistant on the university farm for the following year. This was a complete surprise, for, though I had known that there was such a job each year for someone in their last year, I had not even considered that I might be offered it.

It largely consisted of working on the farm in the mornings, leaving the afternoons and evenings to the requirements of the course, and, in particular, the preparation of the thesis, which involved a good deal of work in the library at the School in Cambridge. I would be paid the very welcome sum of £2 a week. This sounds trivial now, but one pound went a long way in those days, and it paid for my 30 shillings a week digs near the farm, and left ten shillings over, which reduced to some extent the amount my father had to find for my upkeep.

Naturally I jumped at the chance of this job, both for the money, but especially for the opportunity that it provided of getting more experience in the practical field, and of working with a different group of people. It also provided a considerable boost to morale, making me feel, perhaps for the first time, that I might have *some* qualities which could be of value when it came to seeking a job later on.

So, on 1 July 1935, after a short holiday playing golf in Cornwall (something that I actually felt I *was* quite good at!), I embarked on the second stage of my apprenticeship to agriculture.

The University Farm

The university farm at that time consisted of some 750 acres, running westwards from the outskirts of the city in a comparatively narrow strip towards Madingley. Small sections had been detached to form the Plant Breeding Station at the east end, and the Animal Field Station in the middle. The soil varied from quite shallow gravel to heavy and intractable clay at the Madingley end, difficult to handle with the tackle available at that time. It was run as a very conventional mixed farm, with a dairy herd of some 40 dual purpose Shorthorns of good genetic quality, one or two of which were generally entered for the London Dairy Show each year. They were hand milked, and I still remember with affection, Janet, a quiet placid elderly roan lady on whom I learned to hand milk. Bull calves and surplus heifers were kept on and fattened at two and a half to three years of age. There was a sheep flock of 250 Border Leicester/Cheviot ewes, crossed with Suffolk tups, and a Large White pig herd and fattening unit, the sows being kept out of doors on tethers on permanent pasture, and the progeny fattened for bacon. They were marketed through one of the earliest of the co-operative bacon factories in Suffolk.

Virtually the whole range of crops were grown – wheat, oats, barley, and even rye one year; potatoes, sugar beet, kale, beans, one year red clover leys, and temporary three to four year leys. There were basically two rotations. A modified Norfolk four course on the lighter land comprising potatoes, beet, fodder roots and kale, followed by barley undersown with clover for a one year ley ploughed out for wheat. Then on the clay, the rotation was a three to four year ley, ploughed out to a bastard fallow, followed by two wheat crops and barley undersown with the next three year ley. The yields, as I recall them, were 16–20 cwt of wheat/acre (occasionally a little more), potatoes about 5 tons, and sugar beet about 10 tons, and these were typical of those obtained on well run Eastern Counties farms at the time.

This wide range of crops and livestock provided plenty of scope for experimental field trials on topics such as seed rates, fertiliser treatments, times of sowing, prior cultivations and so on, carried on over a period of years, and looked after by successive intakes of students. Many of these trials were written up by Sanders and Garner, and published in the agricultural journals. The trials were not confined to crops only, as there were a number on the animal units, and a major one during my time with which I was closely associated was on the restricted feeding of bacon pigs, and the effects of ad lib feeding on carcase quality.

By today's standards, the farm had a pretty top heavy managerial structure, with Wilfred Mansfield as Director at the top. The next tier down consisted of the Director's Assistant, and beneath him the Farm Bailiff, and at the bottom, the Junior Assistant, and the Office Secretary.

Wilfred Mansfield was a man of strong convictions, with a very practical Eastern Counties farming background. A first impression of conservatism was rather deceptive, for he was, in fact, always prepared to try out new ideas and methods, provided that he thought that they might work. One had the feeling that he regarded the large number of plots scattered round the farm as something of a necessary evil which was a bit of an impediment to his desire to make money out of the farm. I can sympathise with that view to some extent, as it is never easy to integrate the demands of scientific research in an educational institution with a need to demonstrate farming efficiency for the benefit of students.

Having said that, he was certainly progressive in many respects. He was an arable silage enthusiast, and there was an old pre-First World War wooden stave tower silo at the set of buildings opposite Girton College. The story went that one wet year in the '20s, he finished up with a tower full of butyric silage, which smelt so foul that the Mistress of Girton wrote complaining that her girls were completely unable to do any work, and he must do something about it immediately. As he was a Fellow of Emmanuel College and very keen on building up the reputation of the School in the university, he had to empty it quickly and bury the silage somewhere down at the other end of the farm.

Wilfred was a strong advocate of what was then called alternate husbandry (which soon became 'ley farming' under the stimulus of George Stapledon's preaching of its merits) and gave a paper on it at the Third Oxford Farming Conference in 1938. He was a dedicated horse man, having worked with them all his life, and was an enthusiastic supporter of Percherons, which he bred on the farm. This was a result of his experience in the 1914–18 war where clean-legged Percherons gave much less trouble in the Flanders mud than Shires with a lot of 'feather' on their legs. He took to tractors reluctantly, and would argue convincingly on the merits of horses for certain types of farm work.

We did not see much of him on the farm, as he was heavily involved in outside work, but he would generally put in a brief appearance in the morning on his way to teach, or to some committee meeting. He was very closely involved with the setting up of the Land Settlement Association, which had been sponsored by the Government in 1933 to provide smallholdings for unemployed miners. A series of these settlements were set up across the Midlands, providing small groups of horticultural holdings, to each of which was appointed a manager, whose function was to help the tenants on his estate to get established, and to organise the marketing of their produce on a co-operative basis. Each estate had an advisory committee, and Mansfield chaired one of these. There were a good many teething problems, and quite a number of dropouts, and in some cases, I believe, they were sited on unsuitable land. Some of the groups, however, were fairly successful, and it is only quite recently that the LSA was finally wound up, and the holdings sold under the banner of privatisation.

Mansfield was one of the early pioneers of BBC agricultural broadcasting, alongside Arthur Street, James Scott Watson and Anthony Hurd, MP, the influential agricultural editor of *The Times*. Under the leadership of John Green, the first BBC agricultural producer, they played an important part in raising farming standards in advance of the Second World War.

Jem Sanders was also occasionally on the air, but he did not come into real prominence as a broadcaster until the war years, when the service was considerably expanded as a propaganda vehicle for the food production effort, to which he made a distinctive contribution.

I have a small booklet of Mansfield's broadcasts which reflect graphically the changes that were taking place in farming in the later 1930s.

The next tier down in the management structure was the Assistant Farm Director, who had overall responsibility for day-to-day management decisions when Mansfield was away, and he also had the responsibility of keeping all the farm records, ordering supplies, and so on. Under both of them was the Farm Bailiff, responsible for the farm work and the staff, which at that time numbered about 15, including two boys. Then there was also the Secretary in the office, respon-sible for the clerical work and financial records.

So where on earth in this plethora of managerial staff was there a niche for me as Farm Assistant? To some extent, I think that the job had been created mainly to provide one student a year with an opportunity to gain some experience of management, but at the same time to supply a dogsbody to do small routine jobs for which the farm workers were not qualified. The main activities were making weather recordings for the Met Office, which had to be done at 9 a.m. each day, with the results transmitted to London; and secondly, the mixing of rations for all the livestock on the farm. At the start of my year, this was all done by hand with a shovel on the barn floor, and it was a 6.30 a.m. job on at least three mornings a week, and I don't think that I have ever enjoyed breakfast more than in those days. I always had one of the farm staff to help, but by the time we had tipped, turned the heap three or four times with a shovel, and then re-bagged several tons of different mixes for cows, cattle and pigs on an empty stomach, we were ready for bacon and eggs.

This was, of course, before proprietary concentrates had really taken hold of the market, but I think Mansfield and Garner, who compiled the formulas for the rations, would anyhow have opted for farm mixing on economic grounds, and because they liked to know exactly what was going into the feeds.

I have always favoured home mixing on economic grounds on a mixed farm growing a reasonable acreage of cereals, in spite of the grinding and mixing costs. It cannot surely be economic sense to sell cereals at a relatively low price to a merchant, only to buy them back again having carried two sets of transport costs, not to mention mixing costs, share of overheads, and profits of the merchant producing the compound? It may be that my views have been coloured by this early experience of home mixing, but I am sure that a proportion of the profits that I achieved later on when directly involved in farming were partly attributable to it.

It came as a considerable relief in the following spring when the Director decided that the time had come to put in a mechanical food mixer. We were spared the shovelling, but had considerable problems I recall with the cod liver oil, which had to be incorporated in several of the mixes. The injector did not work properly, and we had to do a small mix to incorporate it in meal, and then add that to the main mix. This was, of course, in the days before synthetic vitamin and mineral mixes were available on the market, which only came into general use after the 1939–45 war.

Returning to the management structure, I was once more very fortunate in that Mansfield's assistant at the time was Dick Trehane, later to become Sir Richard, the outstanding Chairman for many years of the MMB (a fourth in the list of knighthoods from Cambridge in that era). I worked very closely with him, as he was virtually resident manager for most of the time in Mansfield's absence. He looked after all the physical records for the farm including the livestock enterprises, and, for some reason, we also did all the castrations of pigs and lambs on the farm, even though there were competent stockmen in charge of both units. In fact, one of our trials at the time was on different methods of castrating lambs, comparing the conventional knife with the recently introduced Burdizzo pincers which crushed the spermatic cord avoiding the necessity for an incision. The trials were followed through to the slaughterhouse with carcase measurements.

We also had running on the farm at that time, the three year experiment on the restricted feeding of bacon pigs – one of the first trials of its kind. This involved a number of paired pigs being taken through to finishing, with one of the pair being allowed to eat as much as it wanted, and its mate being restricted to two-thirds of that intake at its next feed. It involved a great deal of measuring of food consumption, and back weighings of what was left by the ad lib pigs. One of the shortcomings of the trial was that, just as one would find in a sample of the human population, some ad lib pigs were much greedier than others, and in some cases, the two-thirds of a greedy pig's intake, actually exceeded that of the more abstemious pig in the ad lib group. I had the job of setting up the pigs for the second year of the trial and, unfortunately, had not enough weaners from which to find reasonably equal pairs. All my seniors were away at the time, so I could not get advice, and my pairings subsequently turned out to be statistically inappropriate for some reason when it came to working out the results. I am a bit hazy now as to the reason, but I believe it was that I had two pairs of mixed sex, when they should have been unisex. Even so, there were some very interesting results from the trial, which ran concurrently with John Hammond's classic experiments on the nutrition of the bacon pig and its influence on carcase shape and composition.

Though he had only left Reading the previous year, Dick Trehane already stood out as a most mature, confident individual, with a very wide knowledge of agricultural affairs in general, and of Friesian dairy cows in particular. He had obviously lived farming all his life, and his father had been one of the early converts to the Friesian breed with a high yielding herd in Dorset. Dick was then a walking encyclopaedia on Friesians, knowing the history, it seemed, of every animal in the herd book. He rather obviously found his Director's dedication to Shorthorns difficult to bear, but had no luck at all in trying to make the old leopard change his spots – or rather, the spots on his cows.

Dick had already established a close relationship with John Hammond and Arthur Walton, who were then at an advanced stage in their research into artificial insemination at the Animal Field Station on the farm. They were working especially on methods of collection and storage of bull semen. Walton had previously worked with another famous Cambridge reproductive physiologist Professor T. B. Marshall, to whom much credit is due for the earlier research which led up to AI becoming a revolution in animal breeding methods.

I have always felt that Walton received insufficient recognition for the part that he played in the development of AI, but that is often the lot of the back-room research worker.

One day in the spring of 1936, Dick Trehane asked if I was free to give him a hand with a job he was doing for John Hammond. They had been looking for a simple method of collecting semen for research purposes, and they had already developed an artificial vagina with a warm water jacket. They had now hit on the idea of constructing an artificial cow into which this could be inserted. Accordingly they had got a local blacksmith to construct a massive iron frame, of the approximate shape and dimensions of a dairy cow, and Dick had been down to the local abattoir and procured a hide from a slaughtered cow.

Our job was to pad the frame with straw, cover it with the hide and stitch it on tightly, and generally make it look attractive enough to fool the bull into thinking that it was the real thing.

The person who was to collect the semen then crept in under the frame and held the artificial vagina in place while the bull was led out to do his stuff, with the collector no doubt offering up a prayer meantime that the frame would not collapse over him with the bull on top. Sadly, I cannot recall who was the first brave man to undertake the job, but I do remember the bull being led out, and after one false start when it looked as if he might have smelt a rat, he jumped our cow quite happily, and all was well. The cow was used for quite a time, until the technique of manual collection using a teaser cow was perfected. There is a short sequel to this story: some 12 years later, I took a party of my Oxford students over to Cambridge for a four day study tour, during which we paid a visit to John Hammond's Field Station. There, standing in the yard was our old cow, looking slightly weatherbeaten by then, but there presumably as an historical monument to the early days of AI. Walking proudly up to her, I patted her on the rump, and much to the delight of my coarser minded students, uttered the not to be forgotten words 'I stuffed her'.

I learned a great deal from that final year on the farm, not only the practical details of crop and animal production and the day-to-day organisation of the farm work on a fairly complex mixed farm, but also about management and personal relationships from working with the farm staff. I also picked up a tremendous amount of general knowledge about the agricultural industry and its organisations through listening to Mansfield and Dick Trehane discussing matters of policy, or simply talking farming politics, or about developments in the industry. As with Sanders and Garner, I do owe those two a considerable debt for giving me such a good grounding in farming affairs.

All through the course, there were other opportunities for learning about the industry. The student Agricultural Society held fortnightly meetings, at which prominent farmers and agriculturalists spoke on practical, economic or political topics (I recall Arthur Hosier and George Stapledon, for example), and we also had weekly seminars for the diploma students which were occasionally addressed by research scientists of distinction.

But the main purpose of the seminars was to give members of the diploma course the experience of preparing and delivering a paper to an audience. Each one had to do this, either on his experiment on the farm, supported by additional information on that subject, or on another subject in which they were interested. This was excellent training, as I had never before had to stand up and deliver a talk in public, or even to prepare one.

It was quite a daunting experience and I was very glad to have had the experience when it came to the first few lectures in the new job, the following year.

Increasingly as the second year of the diploma progressed one's mind became more and more dominated by the thought of what was going to happen at the end of June if no job had materialised by then, which became ever more likely as time went by. Though the industry was beginning to emerge from its long depression, this did not seem to be reflected in the job market. At that time, there were four types of opportunity for agricultural graduates, outside the purely practical sector. This was long before the advent of the National Advisory Service, which was not formed until 1946, but some of the more enlightened county councils financed small advisory services for their farmers under a County Organiser. In addition, there were regional advisory specialists who mostly worked under university Departments of Agriculture, such as Reading, Cambridge, Leeds, or Newcastle, or at one of the larger colleges. The services of the specialists were utilised by the County Organisers for their more intractable problems, and regional conferences were held from time to time to help to keep those in advisory work up to date with scientific developments.

Most of the County Organisers, of whom there were no more than 20 in the country, had assistants, whose jobs included soil sampling, working out feed rations, preparing drainage plans, organising evening lectures, and so on. They were quite interesting jobs, and a good way of getting a foot on the ladder. The reputations of some of the organisers as people to work for were better than others, and one rather hoped that jobs would not be advertised in the counties where the organiser had a black mark.

Then there were possible jobs in the educational field, either in the universities, the four major agricultural colleges, or in the 12 county Farm Institutes which existed at that time. Jobs at the latter were unlikely to be available to me, as the major qualification here was a very high degree of competence in practical farm skills, which I had not got. I was, in fact, turned down for one job on those grounds, and rightly so, I think. The other two possible sources of employment were research jobs in either government institutions, universities, or the larger commercial companies, such as ICI, or alternatively, sales type jobs with commercial firms. Some of the latter, if they also included advisory work, were not too bad, and a good adviser with a well thought of company could build up a very good reputation in the field and have really a very pleasant job – as long as his sales figures held up, which they would, if he was a respected adviser.

It was rather ironic that by the time I was job hunting, ICI had recently undergone one of its periodic cutting back phases, and had virtually stopped recruiting except for a few very specialist posts, and though my father pulled a string or two to get me an interview with them, no job was forthcoming.

Only two jobs came up in the county field that summer, one as an assistant in Oxfordshire, and one as a county lecturer in Devon. I was short-listed for both, but, in each case, they went to slightly older people than I, and, in the case of Devon, to a close Cambridge friend who had stayed on and done some postgraduate research. That was a most intimidating interview for an inexperienced youngster like myself, as it was before the whole of the Devon County Agricultural Committee, which appeared to consist of at least 30 or more of the toughest and most bucolic looking farmers that I have ever seen, ranged round in a half circle in front of the hapless interviewee, so that one felt rather like a half finished bullock put into a fatstock sale ring by mistake.

Relentlessly the spring and early summer of 1936 moved on, the days filled by farm work in the morning, and the afternoons spent in the library writing up my thesis on the nitrogenous manuring of grassland, though this had to be delayed till after the first cut on my plots. Consequently, as with most academic theses, there was a last minute frantic rush to get it completed in time. There were still no jobs in sight, but the immediate hurdle was the final examination, so the job problem did not figure prominently in one's thoughts for a month or two.

There were three components to the examination in addition to the thesis: set written papers, a farm assessment, and orals with external and internal examiners. For the farm assessment we were taken out to an unknown farm at 10 a.m., provided with essential details and a plan, and then collected again at 3.30 p.m., given a period in which we could ask as many questions as we wished about the farm, and instructed to hand in a full report and recommendations by ten o'clock the following day. The farm on this occasion belonged to one of the early pioneers of arable stockless farming, and I remember that there were plenty of comments to be made about weed control and fertility levels. This sort of exercise, though part and parcel of all management courses today, was well in advance of its time in 1936, and was an excellent experience for anyone likely to be going into advisory work at the end of the course.

Though the practical orals only lasted some 30 minutes each, I like to think that they were of tremendous value to me in my later capacity as Chief Assessor to the National Certificate in Agriculture Examinations Board, in which role I must, in all, have oralled some 3–4,000 students during my years with the Board.

The two examiners were James Scott Watson, then Professor of Rural Economy at Oxford, and Jem Sanders. Scott Watson was a really kind, gentle, and most delightful man, who immediately put a nervous student at ease through his laid back approach, and his apparent genuine interest in the interviewee. He gave the impression that he wanted to know all about you and what you hoped to do, and that he was there to find out what you knew in a positive way, and not to catch you out on what you didn't know. Sanders had much the same easy-going approach. Their attitude made a lasting impression on me and I learned from them the lesson that one will never get the best out of an individual if he, or she, is fearful and ill at ease. My own most rewarding experience of oralling was when a candidate, at the end of a 20 minute oral, got up and shook me warmly by the hand, saying 'Surely I haven't got to go yet, have I? I *was* enjoying our discussion.' Conversely, my least rewarding experience was when, oralling girls on an agriculture and home economics course, I received a note from the course tutor in the middle of the afternoon requesting me not to ask the girls any questions about grassland, as they hadn't covered it, and they were coming out of the orals in tears. At the subsequent examiners meeting, I expressed regret at having caused such distress (they had shown no signs of it), but suggested that the sooner the college included grassland in an agricultural course, the better, and I would expect the students to be able to answer questions on it next year.

Time was up at the end of June, when exams were over (I was duly awarded the diploma), and my job as farm assistant ended so as to allow a student from the following year to take it over. But still no suitable jobs had been advertised for some time, and it was clear that it would be several months before anything permanent could be expected. I have much sympathy for many of today's students who finish college courses with no jobs available, and intense competition for those that do arise. At least I was spared too much competition, as there were not many agricultural students coming out of the universities then, and I had the edge over the ordinary degree course students from having done two years of postgraduate work on what was generally regarded as a prestigious course. But finishing a four years' university career, which had cost my father a lot of money, without a job or any prospect of becoming self-supporting in the immediate future was a somewhat depressing and worrying situation. It was only relieved to some extent by the encouragement provided by

Jem Sanders and Frank Garner, who assured me that something was sure to turn up before too long, and that I could
count on their support, if it did.

So ended the first stage of my apprenticeship to agriculture,
and what a superb apprenticeship it had been, through the
medium of working under three or four of the most outstanding agriculturalists of the day. I had absorbed from them
not only a tremendous amount of factual knowledge about
farming and the agricultural industry, but I had also been
made to think, to question, to analyse, and to commit
thought to paper in a logical, constructive and scientific way.
But above all perhaps I had been transformed from a rather
immature individual lacking any real sense of direction to
someone who had a firm idea of where he wanted to go, and
confident through the knowledge that he had acquired, of
being able to do a reasonably good job somewhere in the
agricultural industry, if given the chance.

But I was still very conscious of an important deficiency in
my apprenticeship so far which could become a liability in
certain types of job. This was the lack of real experience of
doing many jobs in the field, such as working with horses
(still quite important then), tractor driving, and so on. I
decided, therefore, that while I was waiting for that elusive
job to materialise, my training should be completed as far as
possible by getting a farm job for the rest of the summer, or
even longer, if necessary. But getting temporary farm jobs
wasn't too easy in the '30s, as most farmers employed an
adequate full-time staff for their needs, and certainly did not
want to take on anyone extra, because of the Depression.
There were, of course, some farmers who took on pupils, but
only on the payment of a considerable premium, and I didn't
want to commit myself to that, either financially or because I
might want to leave at any time, if a suitable job came along.

A FARM IN CORNWALL

By good fortune, something did materialise straightaway. My
family had moved to Cornwall permanently from Essex

when my father retired in 1926. One of my sisters had become an enthusiastic clay pigeon shooter and was friendly with a farmer's wife from near Newquay, who was also addicted to the sport. Between them, they put considerable pressure on her husband, who really wasn't at all keen on the idea, to take me on. The arrangement was that I should work for them for the rest of the summer, living in with them for my keep. I would have Sundays off to give them a break from having me around. This didn't matter much, as no farm work was done on Sundays in Cornwall in those days, except for essential livestock tasks.

They were a strangely assorted couple, both in their mid 50s, who had married comparatively late in life and had no children. George came from a long established farming family in Cornwall, and like so many other farmers at that time had left school early to come back to work with his father, and had learned much of what he knew from him. But he was extremely shrewd and tough, and though very conservative in some ways, he had a sound business sense, and was always willing to try something new if he thought it might pay off. As the owner of a very good 300 acre mid Cornwall mixed farm he had continued to make money throughout the Depression by good sound farming techniques. He was a man of relatively few words, which were spoken in a gruff broad Cornish accent, which at first sight gave an impression almost of surliness (for which he had something of a local reputation, I believe). But when one got to know him well, one realised that this was something of a pose which he used particularly for his business dealings, and underneath it he was a very kindly and considerate man.

His wife Nellie was completely different. She came from a Devon farming background, was extrovert, and enjoyed an active social life outside the farm. But, as is often the case with opposites, she and George appeared to get on very well, readily making allowances for each other's peculiarities, and I felt very comfortable living with them.

The farm was run on a conventional West Country system for the time, being roughly half grass and half arable, much of the grass being in leys run with the arable on what was then termed an alternate husbandry or ley farming system. There were also a few small, rather wet fields which were in permanent pasture. The arable, cropped with wheat, barley, oats, a red clover short term ley, mangolds and kale, and one field of turnips for the sheep, was very well farmed indeed, being mostly clean, apart from thistles and a certain amount of charlock. This was still some eight years before the first of the systemic herbicides began to appear on the market, and cleanliness was achieved through horse hoeing of the root break, good cultivations, and the rotational grass/arable sequence. Thistles, and, to a lesser extent, docks were a problem as they survived the three year break in grass, and I remember to this day, the torture of handling sheaves of corn, liberally impregnated with thistle stems, until I was able to procure some proper gloves.

On the livestock side, there was a motley collection of some two dozen dairy cows, mostly Shorthorn or Shorthorn/N. Devon cross, with a couple of Friesians, and three Guernseys thrown in to increase the fat and the colour of the milk, which went, via the quite recently established Milk Marketing Board, into Newquay for the summer holiday market. Nellie made some butter and Cornish cream for the house. The cows were hand milked in a number of boxes round a central sunken area, and there was no proper cowshed. The dung was just scraped out by hand, and chucked into the pit and left for months. The milk was carried in the bucket to a rudimentary dairy, where it was cooled over a water cooler and put into the churns. This set-up was very common on many farms, not only in the West, but on small farms right across the country. The dairy herd, if it could be called that, was really only there rather on sufferance, as a provider of a regular monthly cheque, which helped to pay the wages of the four regular workers on the farm, who at the time were earning only about 35 shillings (£1.75) a week. George's real interest was in a small suckler herd of mixed origin, put to a Hereford bull, and the sheep flock of some 150 Devon Longwools which were put to Suffolk tups. His other main livestock interest was his four Shire brood mares, which with a couple of Shire geldings made up the power force for the farm, alongside a reasonably new Fordson tractor. There were also three Large White sows, mainly, I suspect, because there were three pigsties, and it was a pity not to use them! The weaners were sold off in the local market at eight weeks, which was the commonly accepted weaning date in all pig herds.

Looking back now, it surprises me that George who was such a good farmer in all other respects should have been satisfied with such an obviously inefficient dairy set-up. I suppose the reason was that he was basically a beef and sheep man by upbringing, and he had only gone into cows, as thousands like him on smallish farms in the West and in upland areas, when the Milk Board put some sort of a bottom in the market. It guaranteed a reliable monthly income, even if it was not a particularly large one.

Perhaps, though, the milking system was not so inefficient in financial terms as it might have appeared, since the costs were minimal in terms of capital spent on buildings, and the milking was done by the farm workers, so there were no high wage costs involved in specialised herdsmen. The only capital cost had been the provision of the small dairy and cooler, and virtually no money was spent on cowshed washing, or luxuries of that kind. Most of the milk was produced in the summer off grass, so again there were no heavy feed costs to be met either. From the aspect of hygiene, of course, it was hopelessly inadequate by present day standards, and it was not long before George was put under pressure to carry out radical improvements to bring it into line with revised dairy hygiene regulations. This was a situation faced by many thousands of small farmers over the next 30 years, which resulted in the steady decline in the number of farms producing milk, since many decided that the capital expense of improvement would not be justified.

Anyhow, it was a highly interesting contrast to the advanced ideas on strict rationing and hygiene control with which I had been indoctrinated at Cambridge, though I think it taught me the lesson that in farming there are always many different ways of achieving objectives, and that these may often have to be tailored to the local conditions and constraints on the individual farm. Certainly there are basic principles which must be applied, but there has to be a great deal of flexibility in the way in which they are implemented under a wide range of farming conditions. This was a lesson that I was to find of value, when subsequently involved in advisory work during the War Years.

I am sure that, at first, George had very grave doubts about my willingness and ability to work, and probably also felt rather suspicious that my college training would make me critical of his methods. Those were still the days when traditional farmers who had had no agricultural education or training themselves tended to regard 'college boys' as rather cocky know-alls with inflated theoretical ideas which they tried to impose on their elders and betters, which would never work in practice when it came to applying them on the farm! And it must be said that there was a certain amount of truth in that assumption as there certainly was a tendency for those who had been away at college and had seen new techniques and the exploitation of new ideas and come back to a conservatively run farm, to get very frustrated when their suggestions were not put into practice.

So my first week on the farm was quite a tough time, when my new employer set out to test my stamina and determination. It was also a time, I suspect, when he was determined to show, in no uncertain terms, that he was the boss of the outfit, and he was not going to stand any nonsense from a college boy with a Cambridge accent – if there is such a thing.

I had learned to hand milk at Cambridge (not an achievement normally associated with that august institution!), but was not a competent milker, so I was given four cows to deal with on the first morning. One was an elderly Shorthorn who clearly did not take at all kindly to my amateurish assault on her teats, as she resolutely refused to let down her milk till I had struggled with her for some ten minutes, during which time both she and I became increasingly frustrated with one another. Eventually she gave in, but only to the extent of giving me about 50 per cent of her normal offering. That cow never did take to me, though after a few days we came to a working relationship whereby after an initial protest, she would agree to give me a reasonable volume of milk, though never as much, I suspect, as she would have given her regular milker.

The next two in the box were a couple of quite elderly crossbred cows of dubious parentage, who were far more amenable so far as letting down their milk was concerned. The only problem with them was that they had not got much to give, which was perhaps understandable as they had, by the look of them, quite a high proportion of North Devon blood in their veins which would not lead one to anticipate significant milk yields. The fourth component of the box was a second calver pure bred Guernsey, and the problem with her was the size of her teats which were miniscule compared to those of the Shorthorns who had provided the quite generous equipment on which my previous experience had been based. She, again, did not take at all kindly to my inexperienced fumbling with forefinger and thumb which was about as much of my hand as I was able to get round her teats. After two mornings of struggle, during which I had got virtually nowhere with her, apart from several attempts on her part to knock me off the milking stool, I managed to exchange her with one of the other milkers who knew how to handle her, for a more liberally proportioned and docile Shorthorn who was not unlike the favourite Janet of my Cambridge days. Over the next few weeks, I did get a good deal more proficient and speedy as the cows and I got more attuned to each other's idiosyncrasies, but I am sure that they must have all breathed a sigh of relief when I left the farm, and they were restored to their regular milking partners. There was one satisfactory spin-off from these early morning muscular exercises, for when I went home, they seemed to have added an extra 20 yards off the tee on the golf course.

After breakfast on the first morning came the initial test of character with my new employer, which was related to the already mentioned question of hygiene in the milking areas. He had been notified that a visit might be expected later in the week from some official to have a look at the dairy and milking premises, so he thought that he had better do a bit of window dressing beforehand. So I was given a ladder and a brush and put onto sweeping the cobwebs, bird droppings, dust and dirt that must have accumulated over quite a number of years on the roofs and beams (no false ceilings in those days) of the milking boxes and small cooling room.

I thought that I had been getting on quite well by the time George turned up later in the morning to inspect progress. I remember very clearly that I was up the ladder trying to reach some cobwebs which were difficult to get at between two beams.

'What have you been up to all the morning? You should have finished this box long ago,' he growled. 'And what about all those cobwebs up there, I thought I told you to get the roofs properly cleaned down. What sort of a job do you call that?' – pointing at the cobwebs that I couldn't reach. With great difficulty I resisted the temptation of saying 'If you don't like it, come up here and do it yourself', or some other equally retaliatory remark, and explained mildly that I was aware of the bit that I had had to leave, but that the ladder was too short for me to reach it. I had the good sense to realise almost instinctively that he was tempting me to have an argument with him so that he could put me in my place. I did say that I was getting on as fast as I could, but that there *was* rather a lot of mess as it obviously had not been done for some time.

He was less provocative after that, but kept me on cleaning and whitewashing for another two days, so that by the time the dairy inspector came at the end of the week, the place was looking really quite respectable, especially as some of the other farm staff had been put onto cleaning out all the accumulated dung from the central pit between the buildings. As a result he was given a clean bill of health after the inspection.

There were four regular men on the farm; a working foreman who did most of the jobs with the Fordson tractor, and who was a very steady, sound, and dependable man, a carter who was responsible for the shire horse team under the eagle eye of the boss, a general farm worker who was one of the milkers, and a lad of about my own age, who also milked the cows. The latter would not have won a prize for intellect, but was a most cheerful soul who was generally singing or crooning the latest popular song he had picked up on the radio. Three of them spoke in the broadest of Cornish accents, and for some time I had the greatest difficulty in understanding half of what they said, though after a month on the farm I could detect definite signs of Cornish creeping into my own speech. Without exception they were all extremely helpful and considerate towards my own practical inadequacies and willing to help me with things I had not tackled before. At times it was almost embarrassing the way that they would try to spare me the heavier jobs, as if I was some sort of gentleman who shouldn't be expected to soil my hands with hard manual work – which of course was exactly what I was there to do. George himself had a rather bad back, and kept away from the really hard jobs, acting as shepherd most of the time, though the others were perfectly capable of doing all the routine jobs with the flock, as they were capable of doing all the other jobs on the farm. Most farm workers in those days were general workers who could turn their hand to anything – and very competent they were too – with both crop and livestock jobs. Only on the larger farms with enterprises of some size was there any degree of specialisation in the workforce with higher wages paid for specialised skills.

I had arrived just as haymaking had started, and this was, and still is, a tricky business on West Country farms with a fairly high rainfall. Wet weather was one of the reasons why I had been relegated to cleaning up the cowsheds, with haymaking brought to a halt for some days. So till the weather cleared, the next testing job I was put onto was leading one of the Shires horse hoeing the root fields, which were partly in kale and mangolds for the cattle, and partly in turnips and rape for the sheep flock. The root crops were a most important component of the farming system, since they provided the major opportunity for controlling the weeds before the days of systemic herbicides. Just how necessary it was to have someone to lead the horse I am not too sure, for a good horseman with a well trained horse could do a pretty effective job provided that the drilling had been done well in the first place. I've an idea that George was still testing my character to see how it would stand up to a full day's seemingly endless plodding up and down the rows in a field of wet kale. But that, at any rate, was preferable to the next task, which was the singling of the mangolds, as it was at least a job which could be done with a straight back.

I think that hand singling of root crops was about the most boring and tiring of all the jobs that had to be done in those days of labour intensive farming – more tedious even than just hand hoeing of the weeds. The only merit was that one was usually in a small gang, which provided an opportunity for conversation provided that one could keep up with the others, which, to begin with, I couldn't, as I was probably overcautious about cutting out too many plants.

If the weather was reasonably fine, hoeing and singling was usually confined to the mornings, filling in time between the early stock tasks and the time that the hay was fit to deal with after the dinner break. The foreman and the carter would generally have been out windrowing and tedding and turning during the morning to prepare the hay for moving in the afternoon.

Before the introduction of the baler, haymaking was generally a matter of mobilising all hands to the pumps, since it usually involved a five to six man operation either sweeping it up to be built into a rick in one corner of the field, or forking it onto an elevator behind a wagon, or directly up onto a wagon for transporting back to the Dutch barn at the farm buildings. My job was either forking hay onto the elevator at the rick after it had been brought together with a horse-drawn hay sweep, or onto the elevator behind the wagon. The ease with which this could be done depended very much on the skill of the carter in collecting the swath onto the sweep without rolling it together into a tangled rope-like mass. This could be very difficult to tease out to get a decent forkful into the elevator hopper, and it was a race against time to get it done before the next sweep load arrived to cover it all up again, making it still more difficult to unravel. Whether we were carting to the barn or stacking in the field it was all hard labour, and the best moment of the day was when Nellie or the girl who helped in the house arrived with tea and buns – quite often saffron buns, to which I became very partial.

Balers were just being introduced onto farms in the mid '30s, and they were mostly stationary in the field, with the hay being swept up to them as if to a rick. The bales were wire tied, and someone had to stand by to thread the wires through slots in the side of the bale chamber. George, being quite progressive in outlook, decided to hire one of these, and to use it for the hay from the one year ley field of rye grass and red clover. Unfortunately someone had told him that you could bale hay with a rather higher moisture content than if you were putting it into a conventional rick. The advice was that you could safely bale it a day before you would rick it. So this he did, and to this day, some sixty years on, I recall with horror the struggle that we had in trying to manhandle those bales. Even if they had not felt as if they weighed a ton apiece, being wire tied they were terribly difficult to get hold of, and they needed two people, one at each end, to shift them. After the first few, George decided to reduce the density, but even then, they were devils to handle. Fortunately for us, he decided to make a stack of them in the middle of the field where the wind could blow through them and help to dry them out a bit more before they were carted home to the barn. I've often wondered just what they were like when they were opened up in the winter – full of mildew, I would expect.

George, as I have said, acted as shepherd, and I spent a lot of time with him on the flock and picked up a great deal of good, sound practical knowledge in the process. He was dedicated to his Devon Longwools, which were the traditional breed for the area, and he either bred them pure for flock replacement or the sale of ewe lambs to other farmers in the district, or put them to a Suffolk tup for fat lamb production. The disadvantage of the Longwools was their relatively low prolificacy – I doubt if he averaged more than about 120 per cent – and also their size which tended to reduce stocking rate as they were heavy on the pastures in wet weather. But the advantage that seemed to weigh with him was their ability to lamb early and also the size of the lambs when he came to sell them fat – and they were fat, too! I used to try to convince him that he might be better to have a smaller type of crossbred ewe, so that he could keep more of them and have much higher prolificacy, but he would not be swayed from his allegiance, and, I suppose, with a self-contained flock he may have been right for his circumstances and market.

The ewes had already been shorn by the time I came to work on the farm, but the lambs had not all been sold, and their thick fleece and the re-growth of wool on the ewes seemed to act as a magnet for blowflies, which were endemic on the low-lying, thickly hedged permanent pasture fields on which the ewes and lambs spent the summer. Of all the jobs on the farm that I disliked most – and there were not many – dealing with a bad case of fly strike was quite the worst. However careful we were in shepherding – and we were out at least twice a day looking for the telltale signs of a twitching tail or an uneasy or isolated ewe or lamb – one would still come upon the seething mass of maggots hidden beneath the fleece, which until I became hardened to it would give a feeling of absolute nausea. We never went out without the shears and the bottle of Jeyes fluid which was the standard remedy at the time. There was also the occasional case of footrot, but considering the dampness of the pastures this was not as bad as might have been expected, since George had kept it pretty tightly under control.

Dipping against sheep scab was compulsory twice a year in the county at the time, and the early summer dip following shearing came round soon after I came to the farm. This was before the days of DDT and Dieldrin, and the main insecticidal ingredient was arsenic, which was quite reasonably effective in the short term, but gave no lasting protection against fly strike or other parasites as the later dips were to do – until their regrettable demise due to their accumulation in the food chain. It seems to have been exceptionally difficult for scientists to find effective and lasting dip ingredients which are both efficient insecticides and safe for the users, since at the time of writing this the successors of DDT, the organophosphorus compounds, are now under a cloud as well. Arsenical sheep dips, incidentally, had been used on a number of occasions as a vehicle for the final removal of unwanted spouses in country areas, and there was a very famous case not so far away in Cornwall in which Sir Norman Birkett, before his elevation to the Bench, achieved one of his most celebrated acquittals.

Sheep dipping was quite an occasion on the farm, with the local constabulary having to be informed in advance, and the village bobby riding up on his bicycle to see that the job was being done properly. I can't recall that we took any very special precautions against the dip, apart from wearing overalls and wellingtons, which were put on to keep dry rather than as a protection. On the subject of sheep dipping, I have always felt that the Ministry has much to answer for in allowing sheep scab to become an endemic and wide-spread problem again after it had been successfully eliminated by double dipping in the at-risk areas in the '50s. When it did reappear some 10–15 years later, reputedly imported from Ireland, the Ministry should surely have stamped on it immediately by bringing in the most stringent dipping regulations that they could devise, especially dipping twice a year, to prevent its further spread. Of course, farmers would have groused about it at the time, but it would have saved a lot of time and trouble at the end of the day.

I must have passed the tests of willingness to work that George had set me fairly quickly, as he began to open up a lot more when we were working together with the sheep, and the growl disappeared from his voice. He appreciated that I really was anxious to learn as much as I could from him and he became as communicative as his natural reserve would allow him to be. He had never been in this sort of situation before, and he began to quite enjoy, I think, passing on his knowledge, and even discussing other matters relating to the management of the farm. It was not long before he was prepared to send me out on my own for a morning or evening flock inspection, though there were strict instructions to report back if anything was amiss. I took this as a sign that I had now 'arrived', and was accepted as a responsible member of the farm staff.

I never did get an opportunity to work with the horses, except for being sent out to bring them in from the field in the morning, or to take them back to their pasture for the night. One of the mares was a thoroughly bad-tempered brute who would resist all attempts to halter her up in preparation for the day's work if she didn't feel like it. She nearly put paid to my future activities one windy morning when she was clearly feeling particularly obtuse, by swinging round, rearing up and lashing out with her hind legs just as I got the halter on her. Fortunately I just managed to jump clear of her hooves, as it would have been no joke to have been caught amidships by those soup plates with a ton weight behind them. The other horses were much more placid, particularly the gelding, with whom I had established a close relationship in the root fields, in which we had spent many hours together. It didn't really matter very much that I was not able to get any real experience of working with them because although there were still a very large number of working horses on farms then, tractors were beginning to come in very quickly as they became more reliable, easier to start, and generally more sophisticated. The Fordson more or less reigned supreme, and it was to be ten years before the Ferguson was to make an impression on the market.

I had been working on the farm for some six weeks or more when there was a phone call from home to say that there was an advertisement in *The Times* for an assistant lecturer in agriculture at Reading University, so I wrote off for the details. Before they arrived there was a letter from Jem Sanders at Cambridge also drawing my attention to it. He said that he thought I ought to apply, and if I did so, he and Frank Garner would be pleased to act as my referees. This sounded a marvellous opportunity, for Reading then, as now, had a very good reputation in the agricultural world, and it would obviously be very good to have the support of my tutors from Cambridge. I didn't really think that I would have much chance of getting the job in view of my obvious lack of experience, and, to be honest, I felt considerable trepidation at the thought that I might be appointed, as I was still lacking in self confidence. But it was already at least two months since anything at all suitable had been advertised, so I sent off the application and awaited results.

In due course, a letter arrived back, summoning me for an interview at Reading in a couple of weeks' time. This was timed for 2.30, so I decided to stay overnight with my Cambridge friend in Totnes as it might be a bit risky to try to go all the way up from Newquay by train that day.

I do not know to this day how I managed to get the times of the trains from Totnes to Reading wrong; suffice it to say that I duly turned up at the station the following day to catch the 10.20 (I think it was), only to stand on the platform to watch the train come roaring through at about 70 m.p.h. without stopping.

It soon became apparent that there was no way in which I could possibly get to Reading before 4 o'clock that afternoon. It was about the worst moment of my life to date, and I don't know which emotion was most dominant – regret at having blown the chance of getting a job at last, and a very good job at that, or plain fury at having been so stupid as not to have double checked train times on such an important occasion.

Fortunately, I had the telephone number of the university in my pocket, and even more fortunately, they were not interviewing anyone till the afternoon, so I was able to get straight through to Robert Rae, the Professor of Agriculture. He listened to my abject apologies and tale of woe with great kindness and sympathy. 'Don't worry at all,' he said, 'we're not due to finish the interviews till 4 o'clock; so we'll have a cup of tea, and see you at 4.30 if you can make it by then.' He sounded so nice and pleasant on the phone that I felt slightly better, but still could not believe that they would ever offer me a job under such circumstances.

I arrived with just six minutes to spare, and right on the dot of 4.30 presented myself, hot and dishevelled, for the interview, quite convinced that no sensible group of people would ever wish to employ such a stupid and irresponsible person who could not organise the most important meeting of his life properly.

But I was wrong, for after a most friendly interview chaired by Robert Rae, and an anxious wait of some 20 minutes, I was summoned back to be told that the job of Assistant Lecturer in Crop Husbandry at a salary of £250 a year on a three year term was mine, if I was prepared to accept it. Of course, I was, even if I was panic-stricken at the thought of what might be involved.

The job was to be assistant to David Black, the lecturer in crop husbandry who was taking up the responsibility of setting up and running a new poultry unit at the old university farm at Shinfield. Some 18 months previously the university had purchased the present 350 acre university farm at Sonning, and already the dairy and sheep units had moved over there, leaving room at Shinfield for a large, modern poultry set-up which would be the leading university department in that field. David wanted to shed a lot of his crop work to give time for building up the new unit.

As I was catching the night train back to Cornwall, there was some time to spare, so David took me out to his home on the Sonning farm for a high tea and a quick look round the farm. He was an Edinburgh trained Scot, as was his wife, and in his pram was their recently arrived son, now Dr Murray Black, the Treasurer, among other things, of the British Society of Animal Production. They made me feel very welcome, and David turned out to be a very good friend and supportive colleague. Already I was getting the feeling that I was coming to work in a friendly department, which went some way to allay the considerable alarm that I felt as to my ability to cope with the job that lay ahead – an alarm that seemed to intensify during the early morning hours as the night train rattled back to Cornwall.

George and Nellie were pretty surprised when I told them the news the following morning, as I don't think that they had expected me to be offered the job. I sensed, too, that they were rather disappointed, as I had become almost part of the household, and it would mean that I would be leaving the farm in the near future. I promised to stay on for another three weeks to help with the harvest, which was just starting.

That they were disappointed at the prospect of my leaving was borne out by an astonishing proposal that George put to me the following morning when we were out together going over the ewes.

'Look, Mike,' he said, 'are you quite certain that you want to take up that job in Reading, and that you'll be able to make a success of it? I think that you're much more cut out to be a farmer, and that's what you ought to do. There's a very good 250 acre farm near St Columb, which has been on the market for some time, and it's going cheap. I'd be prepared to take it, and put you into it. I'd set you up there and find all the money, and keep a good eye on you to see that you got on alright. What do you think about it?' George was, in some ways, a very diffident man and slow of speech, and he obviously found it difficult to put his proposal into words and to express his exact feelings. But it was very clear that he was, and I suspect his wife was also, genuinely upset at the thought that I would be moving on out of their lives, and that they really did want to give me a good start in a farming career.

It is possible, too, that they had another thought at the back of their minds in making this proposal. They had no children of their own or close relatives, except George's elderly brother who farmed not far away, and they were both getting on in years. They may have thought that if I got on well I would be able to take on their farm when he wanted to retire in a few year's time. This would ensure a succession, and would mean that he could retain an active interest in the farm, to which he was very attached.

Naturally, I was completely taken aback by this offer, which came right out of the blue – and very moved by it, too. It meant that George and Nellie had come to accept me personally as almost a member of the family, and also that I had reached the farming qualifications set by George. It was a very considerable morale booster to someone who had been a fairly slow starter in life and was still very uncertain as to his capabilities to do a job properly. I thanked George profusely, but said that I would like 24 hours to think about it, though I was already almost certain in my own mind that the answer would have to be no.

Firstly, I had no money of my own, and I would have been unhappy to rely on George's generosity entirely for setting me up in the farm. I felt that I simply could not rely on his money for everything. Secondly, for domestic reasons, I didn't relish the idea of living alone or having a housekeeper. Thirdly, I had already accepted the job at Reading, and did not think that I should turn round now and withdraw from it after the support that I had received from Cambridge. Though the prospect of it still filled me with trepidation, I felt it to be a great challenge from which I should not run away.

The following morning, I told George, as gently as I could, that I really thought that I ought to stick with the Reading decision, as I had already taken it, much as I appreciated the kindness and generosity of his offer. Naturally, I have often wondered how different my life would have been if I had accepted. Would I have succeeded and become some sort of farming tycoon, or a part-time farming politician, or would I have stagnated as a small, not very successful West Country farmer? We shall never know, and that is just as well.

There were to be other crossroads with seductive signposts pointing this way or that in my life, and it is pointless to stop and wonder now what the results of following different roads would have been. It would be particularly pointless, anyhow, as I cannot think that I could possibly have had a better agricultural life than the one that resulted from the turnings that I did take at different times along the way.

So, for the time being it was back to the farm to help with the harvest, and I don't think that any other aspect of farming has changed quite so much with the advent of mechanisation as has harvest and grain handling. It is customary nowadays to look back nostalgically to harvest as a time when all the farm staff worked together in unison in fields of golden corn under a glorious summer sun, stooking sheaves behind the horse-drawn binder, or pitching them onto the wagons to be taken in for building into symmetrical ricks in the fields or the rickyards. Every now and again, the farm wives and children are imagined to appear bearing jugs of hot tea and buns or cake, to revive the flagging energies of the workers. Then, as the binder gradually reduced the area of the standing corn, the farmer and his son and the foreman would pick up guns, and wait for the fleeing rabbits as they decided to make a dash for it from the ever decreasing area of corn. I must confess that I always felt extremely sorry for those poor rabbits having to make for the nearest hedge across a no man's land of stubble covered by gunfire from all directions.

This ritual finish to the cutting of the corn in the field certainly did provide a bit of excitement to an otherwise uninteresting day, for there was really no glamour in spending a day picking up and stooking thistle infested sheaves of corn.

To some extent, of course, this rosy picture of past harvests *does* have an element of reality about it. Harvest *was* looked forward to as a time when all the work that had previously gone into growing the crop came to fruition, and it could become a time for celebration. And it was also one of the few times during the year when all the farm staff worked together in the field, and when the normal routines of milking and shepherding and the feeding of livestock were disrupted in favour of getting a crop set up against the weather in the field as quickly as possible, or carried safely to the barn or rick against the threat of an unfavourable forecast. There was a sense of common effort towards preserving safely the fruits of the previous year's labours. There was also something of a sense of excitement and anticipation in the days beforehand when the binder was brought out of the barn and greased up and the canvases and knotter checked over for the last time.

But harvest certainly wasn't the holiday that present day romantics imagine it to have been. Most of it was sheer hard work and drudgery, and in wetter summers it could be desperately frustrating. If the crop was laid or damp or a bit tangled, the binder would be for ever bunging up, and both horses and men would become testy and short tempered. Even when things went well a day in the field stooking corn could be very exhausting due to the continual stooping down to collect the sheaves and carrying them to a point where there were enough to make a shock with the ears together, and leaning so that any rain would run down the outside and not into the centre.

The amount of hand labour that went into the crop between cutting and the final threshing of the grain was astonishing, quite apart from the horse and engine power involved. Firstly, the sheaves had to be picked up and put into the stook, then, if the weather was catchy, the stooks were often thrown down by hand to dry out a bit more before carting. In very bad weather, the stooks might have to be opened up and turned. They were then pitched up on the wagon and manhandled into place to make a tidy load which wouldn't tip over on its way to the rickyard. Once there the sheaves were pitched from the wagon to the rick builder, with perhaps an intermediary in the middle to pass them if it was to be a sizeable stack. If the stack was outside, and not in a Dutch barn, it would then need to be thatched to protect it till threshing in the late autumn or winter. The sheaves would be pitched by hand from the rick to the man on the thresher, who cut the bonds before feeding them into the machine. At the other end, was a man bagging up the grain, with another stacking the sacks of grain, one more dealing with the straw, and the dirtiest job of the lot, a man dealing with the chaff and cavings. In all, the threshing gang would consist of at least eight men, and a sheaf of corn from the binder would have been manhandled eight times before it was fed into the threshing machine.

Compare all this with the farm of today, where an 18-foot cutterbar combine harvester, with a straw chopper on the back, will go into a field of ripe corn and eat up some 40 acres or more in a day with only one man at the controls. One man will be employed carting the threshed corn back to the drier or grain store, and another overseeing the drier and the storage of the grain in the bins. Just three men dealing with a far larger acreage in a day than the largest gangs of the past could possibly deal with. Even if the straw is not chopped and ploughed in behind the combine, one man can bale some 50 acres of straw a day, and the bales will probably be wrapped, and all handled mechanically. If small balers are still used, some man handling will be required at the farm, but it will be minimal compared with the past.

This must be one of the best examples of the substitution of hand labour by machine power for any industry in the country, but, of course, a heavy price has to be paid for it in mechanisation costs. The investment in harvest machinery and grain storage is extremely high, and the sad thing is that quite a high proportion of the machinery can only be used for a very short time, and must therefore lie idle, earning no return for the farmer, for at least ten months of the year. In this, farming is very different to most other industries where expensive plant will be run for the maximum number of hours in order to cover its purchase price.

One cannot regret the passing of the old order, with its drudgery of hard, continuous and repetitive hand labour, and the resulting physical disabilities that so often followed from it in terms of human health. Perhaps the best thing of all was that mechanisation finally led to the demise of the greatest threat to the health of the farm worker, the 2¼ cwt grain sack, with its toll of hernias and damaged backs.

So after another three weeks helping with the harvest, it was goodbye to the farm, and, sadly, to George and Nellie, of whom I had become very fond and to whom I owed a very great deal. The second stage of my apprenticeship to agriculture had come to an end, and it was time to launch out into the next phase.

But before embarking on a further account of my personal odyssey, this might be an appropriate point at which to stand back and take a brief historical look at the state of the agricultural world that I was about to enter as an active participant.

AGRICULTURAL BACKGROUND 1880–1936

Preface to 1914

It is customary to think of the period between 1880 and the mid 1930s as a time of unrelieved gloom and depression in the agricultural industry, when prices were low and Government support for farming negligible. To some extent this is a true picture, for Britain as a world power with a large colonial empire partially dependent on it, had to be a trading nation. It was a better economic proposition to import cheap food from developing countries and the Americas, and export manufactured goods in return than to support a home farming industry whose prices, because of a higher standard of living in Britain, were uncompetitive compared with those for products coming from the New World. The rot had set in during the last quarter of the 19th century with the development of large, speedy ships with refrigerated holds, which made the shipment of animal products and grain to Britain a feasible proposition, even when transport costs were added. Cheap food kept manufacturing costs down, and a measure of the disastrous effect that this had on farming was that some 45,000 acres of land was lost to agriculture annually between 1890 and the outbreak of the First World War in 1914. A good deal of this went for building, but quite a sizeable area went out of production to revert to scrub and woodland on the more marginal light land soils or the heavy clays, which were very difficult to work when only horse power was available.

But in spite of this general depression, the picture was not entirely negative, for scientific research began to expand, and there were always some pioneers anxious to find new ways of beating the recession by changing methods of production. These had been greatly assisted by the 1908 legislation which gave freedom of cropping to tenants (and at that time 85 per cent of farms were held on tenancies) which permitted farmers to break away from the rigid rotations prescribed by their landlords. It enabled them to exploit new systems of cropping made possible by the greater availability of artificial fertilisers with less dependency on animal manure to maintain fertility of the land. In 1909, Lloyd George as Chancellor of the Exchequer provided an additional impetus to development by getting a vote of £2,500,000 through Parliament for the expansion of countryside resources. Part of this money was devoted to the setting up of research institutes, such as that at Shinfield, which became the NIRD subsequently (National Institute for Research in Dairying).

It was during this period that a new type of agricultural technologist began to appear on the scene, men who were to lay the foundations for the truly tremendous expansion in agricultural productivity that was to come some 40 years later. Somerville and Gilchrist pioneering the use of phosphate to increase grassland production, Daniel Hall and John Russell developing fertilisers at Rothamsted, Rowland Biffen breeding new and higher yielding varieties of wheat, and T. B. Wood bringing a systematic approach to the feeding of livestock were all examples of this new breed of agricultural scientist. It is interesting to speculate why many of these received their knighthoods from a presumably grateful Government, but that that same Government did little else to alleviate the economic depression in the industry.

Then, shortly before the outbreak of war, the first tractors began to make their appearance on the farm. For many years, they were little more than glorified horses, and it was not till the 1930s that they really began to assume much real importance on farms.

The 1914–18 War and Its Aftermath

The Government appears to have entered the war in a singularly complacent mood so far as food production was concerned. They were expecting quite a short war, and were confident that the large British merchant fleet would continue to bring in all the food that was required. It is a terrible indictment of the policies of the previous 40 years that, by 1914, British farms were only producing about 40 per cent of the food consumed in the country. The Government did set up a number of committees in 1915 to make recommendations for increasing food production if it became necessary, but with a good harvest that year, no action was taken.

Then in 1916, there was a bad harvest both in Britain and America (which had been supplying a lot of grain previously), and in 1917 the U-boat campaign intensified alarmingly, with 200,000 tons of shipping being lost in September alone. (I well recollect, hundreds of tins of rotting meat from a torpedoed ship being washed up on the beach when we were holidaying in Cornwall that year.) Quite suddenly the situation had become very serious, with the war hopelessly bogged down in France, and food supplies not getting through. The second coalition Government under Lloyd George was formed in the autumn of 1916, and it immediately went into action, with a new President of the Board of Agriculture and Fisheries (equivalent to the Minister today), Rowland Protheroe, later Lord Ernle. He moved rapidly to implement many of the recommendations of the 1915 Lord Milner committee. Measures were taken to set up a Food Production Department and local War Agricultural Committees to implement policies at ground level. Powers were provided to take over land for producing food, and to give orders to farmers to plough up grassland for the production of cereals and potatoes. The target set was to restore the arable acreage to that of 1870, when farming had reached its zenith, over the next three years. This would have been an impossible task, given the run down state of the industry at the time, as it would have meant bringing back some five million more acres into arable production, and there would not have been the labour or the horse power to do it.

A Corn Production Act was passed that autumn which gave farmers for the first time a guaranteed price for cereals, and it laid down how the assistance was to be given. By now, labour was getting very short on farms, due to the terrible losses in France and an ever older call-up policy, so the Women's Land Army was established, which with prisoner of war labour was just about adequate to deal with the extra acreage in production over the next two years.

Of course, the ploughing up campaign came too late to have any effect on the wheat acreage for 1917, but 975,000 acres were ploughed, going mostly into spring barley and potatoes. By 1918, the extra acreage had reached a figure of three million, though, by then, the worst of the crisis was over. But it was quite a remarkable achievement by farmers, all the same. 1917 was the worst year as far as food supplies were concerned, with sugar the commodity in shortest supply, and other foods tightly rationed. The shortage of sugar turned out to be something of a blessing in disguise for farmers in the post-war years.

The wartime campaign, short as it was, proved that it was possible to increase food production quite rapidly in an emergency, since an increase of 25 per cent in energy foods was achieved over the first two years. But perhaps the most important aspect of the Food Production Campaign was that it provided a blueprint for the type of organisation required to increase production in an emergency, which proved to be of inestimable value when the Second World War brought about a similar crisis 20 years later.

The decade following the end of hostilities in 1918 proved to be one of bitter disillusionment for farmers. The Government that, as late as the autumn of 1920 had confirmed that the guaranteed price system for cereals would be continued, suddenly repealed the Act in the summer of 1921 because of its cost, and replaced it with a once-for-all payment for the 1921 crop. The effect was catastrophic. Within two years, the price of cereals was halved, and many bankruptcies followed, not least among the many ex-servicemen who had invested their gratuities in small farms, and who had not had time to establish themselves before the crash came. Thus was born the bitter distrust of Government assurances among the farming community which persisted right up to the 1947 Act after the Second World War – and even beyond it. To farmers with long memories, the repeal of the Corn Production Act still stands as an extreme example of Government perfidy.

So after a brief honeymoon period when farming had prospered under a benevolent price system, it was back to recession throughout the 1920s, almost worse, from the purely economic aspect, than that which preceded the 1914 war, though the period was noteworthy for two measures which did have a lasting effect. The first was the Sugar Beet Act of 1923 which financed the building of a chain of sugar beet factories, mainly in the Eastern Counties, and gave subsidies to growers while they learned the techniques of growing the crop. This provided a lifeline for farmers in the arable districts who had been the worst hit by the collapse in cereal prices. The Government took this action largely as a result of the scare in 1917, when there was reputedly only ten days supply of sugar left in the country. It was to pay handsome dividends in the 1939–45 war, when the whole of the domestic ration (albeit a pretty small one) was met from home produced sugar.

The other noteworthy action in this period was the derating of farm land and buildings brought in by Winston Churchill as Chancellor of the Exchequer in 1926, which certainly helped farming by removing one of its larger fixed costs.

Prices improved slightly after 1926 and a small degree of confidence was returning when commodity prices slumped again in the world financial crisis of 1929. Britain became a dumping ground for world food surpluses at rock bottom prices, and home cereal prices fell by 34 per cent over the next three years to equal those of 1914, though with considerably higher production costs to bear. At this time, some 100,000 acres a year were going out of production, much of it reverting to scrub and untended grass and rushes on the wetter land. Ditches became silted up, hedges spread out into the fields, and the countryside, especially in those areas of poor and difficult working soils presented a sorry picture of neglect. Landlords found it almost impossible to let vacant farms, and it was not uncommon for them, if they found a suitable tenant, to let a farm rent free for two or three years on condition that the new tenant cleaned it up and brought it back into condition. On a personal note, I have clear recollections of three examples of the effects of the depression, though these date from the early '30s rather than the late '20s. The first is of the road from St Neots to Cambridge with many fields covered with scrub and bushes – land which now grows excellent crops of wheat. The second is the railway line from Didcot westwards to Swindon and Bath with wet, sodden, rush-infested, undergrazed grassland – a typical example of so-called 'dog and stick' farming. The third is of the rabbit-infested chalk downlands growing ragwort and a host of other weeds in Hampshire and Wiltshire on some of

44

the thinner chalk soils, where farms were difficult to let. These were farms which in earlier and more prosperous times were kept in production by flocks of close-folded sheep and good prices for malting barley.

This was, in fact, the dark before the dawn, for the National Government formed in 1931 to extricate Britain from the financial collapse of that year, at last began to reverse the agricultural policies of the past 50 years. It did this in four main ways: regulation of imports, marketing legislation, subsidies for selected commodities, and grants to improve efficiency.

Under the Ottawa agreement, Empire countries agreed to quotas on the amount of their exports provided that they were given preference in the market over other foreign countries, who had to pay import duties. This affected the imports of meat particularly, while international agreements were made in respect of wheat and sugar. This stopped the dumping of foreign food at very low prices.

The Marketing Act of 1931, brought in by Lord Addison just before the national Government was formed, gave powers to producers to set up Marketing Boards for any commodity provided that more than 90 per cent of producers voted in favour of one, and a further Marketing Act in 1933 allowed for the control of imports for any commodity for which a Board had been set up. Milk was the first commodity to obtain a Board, voted in by nearly 100 per cent of the producers. It proved to be the most successful of them all, and to one who can just about recall the chaotic state of the milk market before it was instituted, it seems crazy that it should now be disbanded, primarily because of EC demands. Farmers in the years ahead will, I feel sure, come to regret its passing after 60 years of mainly successful operation. But it seems part of life that everything has to change whether it is serving a worthwhile purpose or not.

In all, 17 Boards were set up in the UK over the next six years, but many of them did not last for long, and those which failed were usually associated with commodities which were imported in large quantities, owing to the difficulty of controlling imports effectively. Those which survived for the next 50 years or so were the four Milk Boards, and those for potatoes, wool, and hops.

The success of the Milk Boards was largely due to the fact that milk is a very perishable commodity, not until recently imported in liquid form, and it was possible to create a monopoly situation to protect home based producers. But the Board also provided an organisation with sufficient strength to bargain with the big distributors and trade bodies, and it gave some stability to the industry, which encouraged farmers to invest capital in modernising their milk production units, and this was to prove the case right through to the introduction of quotas in 1983.

The third form of assistance given to farmers was through subsidies for commodities deemed to be in need of special help where other forms of aid were inappropriate. Two examples of this were sugar beet and beef. The sugar beet industry was completely reorganised in 1936 with the creation of the Sugar Beet Corporation in which the Government held the majority share. This was empowered to control the development of the crop in all its aspects, including the provision of contracts, management of the factories and the marketing of the product. This, like the Milk Board, on the whole served the industry well for the next 40 odd years.

The other form of subsidy introduced at the time was the Deficiency Payment. In its simplest form, a standard price for a particular commodity was set by the Government, and then the actual average price paid in the open market over a certain period was calculated. If this was below the standard price, producers who had sold over that period were reimbursed the difference. This had the advantage of encouraging efficient producers to get high yields, and thereby a better overall return, but put a bottom into the market for those who were only average producers, perhaps because of poorer local conditions. But the standard price was set so that it did not permit really inefficient producers to stay in business.

The system was only suitable for a limited number of products such as cereals, but it was the basis of much of the support given by the Government, both during the Second

World War, and until 1973, when it had to be abandoned on entry to the European Community.

The fourth type of assistance were grants to improve the efficiency of farming. These were direct payments to farmers to help pay for specific farming operations, such as land drainage, application of lime to acid land, or for the ploughing up of old pastures. These were not introduced until 1937, when the shadows of war began to gather on the horizon with the rise of fascism in Italy and Germany. Grants were also provided for national improvement schemes, such as clean milk production, the eradication of disease through the State Veterinary Service, and for education and research.

The cumulative effect of these measures in the 1930s was an increasing growth of confidence among farmers that the Government did not intend to let the industry sink back to the levels of 1929–31, when the home industry was once again only producing about one third of the food consumed, and an area comparable to the size of Wales had disappeared from agriculture compared to the acreage being farmed in 1870. But even at the outbreak of war in 1939, the figure for home produced food was only approximately 36 per cent of total consumption.

This was the position on the economic front throughout the '20s and '30s, but the gloomy picture did not represent entirely what was actually happening on the ground, for in the technical field, tremendous changes were already beginning to take place – changes which were the forerunners of the revolution which was to change the face of agriculture over the next 40 years.

Adversity can often act as a spur to progress, since there will always be a number of progressive operators prepared to try out new ideas and techniques as a means of achieving more efficient production, either through a reduction in labour costs, or through greater output in relation to inputs. This was certainly true of farming in the inter-war years. One of the great innovators of the period was Arthur Hosier of Wexcombe who designed a portable milking shed, or bail, which could be pulled by a tractor across his downland farm in Wiltshire, thus enabling milk, which was one of the more stable products, to be produced from land well away from his buildings. That allowed him to introduce leys into his arable rotation, giving a much better return through milk than he could obtain from beef or sheep, and furthermore, increase the yield from his cereals when the leys were ploughed up owing to better fertility left behind by the dairy cows. The cows were milked by machine in rotation through six standings, and the bail moved on as soon as the surrounding land began to get poached. He was greatly helped by the recent introduction of the portable electric fence which helped him to pen his cows before milking. He developed his system, which was quite widely adopted on extensive farms in the south, in the early 1920s, and it is strange that it took something like another 20 years for the principle of milking cows in rotation through a small number of standings to be adapted for use in permanent buildings on the farm, with its great saving of labour cost, compared with the traditional cowshed, or byre.

Another area in which considerable technical advances were being made was in animal feeding, particularly with dairy cows and pigs. The research of T. B. Wood and H. E. Woodman at Cambridge had led to the formulation of a relatively simple rationing system, employing cheap concentrate feeds, which were mainly a result of the very low cereal prices and low cost imported protein. I well remember Hugh Mattinson, for whom I worked in the early part of the war, telling me that most of his advisory work as County Organiser for Surrey in the inter-war years was related to dairy cow nutrition and management. In the dairy sector, too, milk recording societies had sprung up to encourage the establishment of officially authenticated records for dairy cows, which also served as a useful aid in feeding and management. Add to this, the assistance being given to clean milk production and disease control through tuberculin testing, and there is a picture of rapid technical progress in the dairy industry, underpinned further by the formation of the Milk Marketing Board in 1933, and the greater feeling of stability that this had engendered.

This was also a period of considerable development in

grassland husbandry and the application of plant breeding in an attempt to produce strains of grass with higher yields of leaf and stem, and more persistency. Much of the breeding work was carried out at the Welsh Plant Breeding Station at Aberystwyth under the dynamic leadership of Professor George Stapledon (later Sir George). He was a grassland prophet who preached its merits with Old Testament fervour and he did much to make the farmer aware of its potential if only it was treated with the same degree of expertise that the arable farmer employed in growing cereal or root crops. He concentrated, too, on the merits of what used to be called alternate husbandry, but which he rechristened ley farming. He was to become the first Director of the Grassland Research Station initially at Dodwell in Warwickshire, before its move to Hurley. At the same time, in the early '30s, Martin Jones at the ICI Research Station at Jeallots Hill in Berkshire was developing new controlled grazing techniques, employing much larger quantitites of nitrogenous fertiliser. In this he was, too, greatly helped by the introduction of electric fencing. The outstanding contribution of these two pioneers, ably assisted by their research teams, provided the basis of modern intensive grassland farming.

Mention has already been made in the opening chapter to John Hammond and Arthur Walton and the development of artificial insemination, but this was not the former's only outstanding contribution to the industry. In the '30s he was conducting his trials on the effect of levels of nutrition on the carcase characteristics of both sheep and pigs – studies subsequently extended to beef cattle. These experiments were written up in a number of classic papers in the scientific journals, and his pioneering research into meat quality has provided the basis of much of the improvement that has been achieved over the past 50 years.

But, perhaps from the practising farmer's point of view, one other change taking place in the industry transcended all these technical achievements – progress in mechanisation. Throughout the inter-war years, agriculture was slowly but surely becoming more mechanised, even if this appeared to be having little effect on the working horse population, which remained high, right through to the late 1940s. In 1925, the Government had set up the Institute for Research in Agricultural Engineering at Oxford, which, after a rather chequered start in respect of its first Director who managed to corner a proportion of its funds for his private use, did much useful work in testing and assessing the value of new machines over the next 15 years. It was moved to Yorkshire for the duration of the war, and then metamorphosed into the National Institute at Silsoe.

The first combine harvester had been imported from the United States in 1928, and worked on a farm in Wiltshire. Two more came in for the 1930 harvest and one of these was based on Dick Warburton's farm at Shillingford, near Oxford, which enabled valuable research into its suitability for British farming to be carried out by the scientists at the institute there. Many problems were encountered, for the American machines were designed for dealing with much lighter crops than were generally found in Britain, and they also had to deal here with longer and often much damper straw than in the prairies of America where they had been developed. The combines were also trailed models, and the tractors available at the time were often not really powerful enough to allow the machines to work efficiently in wet soil conditions.

It was to be another 15 years before the self-propelled models, the precursors of the large machines available today, made their appearance from the States. But in the meantime, considerable improvements were made to the earlier prototypes, such as larger threshing drums and straw conveyors, which allowed them to cope better with British conditions.

As so often occurs in farming, the introduction of a new machine or technique brings in its wake a whole new set of problems. In the case of the combine harvester, it meant that grain driers had to be designed to cope with the grain coming off the harvester, which was often too damp for safe storage for any length of time. On the earlier harvesters, the grain had to be bagged up on the combine, and the sacks dropped off in the field, and then manhandled onto trailers, so there was still a considerable labour input involved, though not so much as with the conventional binder and threshing drum.

The straw, too, was left in long windrows in the field and this had to be baled up mechanically as it was far too bulky to be handled efficiently with manual labour. The advent of the combine undoubtedly acted as a spur to baler development, and it was not long before stationary balers using wires to tie the straw (or hay) were replaced by automatic twine tying models, pulled behind the tractor, which left the bales lying in the field. The next development some 20 years later, was the bale collector or sledge, which assembled the bales in groups of eight, which could be picked up and stacked mechanically.

As a result of the great interest building up in the mid '30s in mechanisation as a means both of reducing labour costs, and speeding up farming operations, a group of farmers met at Harper Adams Agricultural College to pool their knowledge on the subject in 1935. From this meeting was born the first Oxford Conference on Mechanisation in Farming, organised by the Institute for Research in Agricultural Engineering and the Institute of Agricultural Economics, which was held in January 1936. This was so successful, that it became an annual event until the outbreak of war in 1939, to be revived in 1951.

This, then, was the state of the agricultural world that I was entering as I left Cornwall in September 1936 for Reading, full of trepidation at the start of my first job. It was an industry still economically in considerable difficulty: many small farmers had a low standard of living and were in financial trouble, but there were enough larger farmers with adequate financial resources available to allow them to experiment, and to exploit many of the new technical developments that were emanating from the chain of research stations which had been established over the previous 30 years, and from a number of progressive commercial firms serving the industry. It was an exciting time.

CHAPTER TWO

First Practice

———✺———

THE READING YEARS*

WHEN I arrived at Reading in 1936, it was still a small university of some 700 undergraduates, of whom about a third were on courses related to agriculture, horticulture or dairying, so, unlike other universities, agriculture had a very powerful influence. The agricultural department had grown from part-time courses instituted by Reading Corporation in 1893. These were soon grouped with other courses to form an Extension College, which was finally granted full university status in 1926. It is interesting that several Oxford colleges were involved in sponsoring the early extension schemes, including what was to be in the future my own college, Christ Church.

The university was based on its own rather small campus on the London Road, adjacent to the Royal Berkshire Hospital. It was a very pleasant area, with a good central library, assembly hall and Senior Common Room, with other departments reasonably well spaced out. Apart from agriculture, there was a good range of other disciplines stretching across both the arts and the sciences. Because everything was on a small scale, the Senior Common Room was a very friendly place. Undergraduates were mostly housed in five

* A few impressions of Reading at the time have already been recorded in *The Silent Fields*, the 1993 centenary account of the Agricultural Department at the University. I am much indebted to Dr Paul Harris for permission to re-use some of that material.

Halls of Residence outside the main campus, three of these were for women, and the other two for men, and for the period, the ratio of women to men in the student body as a whole, was probably higher than in most other universities.

In the agricultural department, we were very short of room, and I was given a desk in the room occupied by the Head of the Machinery Department, which was housed in an old army hut. A Major from the First World War, J. B. Passmore was a colourful character who had the most remarkable range of risqué and scurrilous stories of anyone I have ever met, which were produced on the slightest provocation. He doubled as the Warden of the largest of the men's halls – Wantage, and dining at High Table was never dull! I did not see a great deal of him in his office, as he was due to retire that year and was seldom in. But he must have sized me up favourably, because after the first term he offered me one of the staff flats in the hall, which I very gladly accepted, as it was far more comfortable and convenient than digs, with all meals provided in hall for a very low charge. The only official responsibility was to read prayers before breakfast twice a week, which wasn't a particularly onerous task. The reading of prayers had been one of the conditions laid down by Lady Wantage, the widow of the Crimean War VC, Lord Wantage of the Lockinge Estate, when she donated the money to build the Hall in his memory in the early 1900s.

Robert Rae had been appointed to the Chair of Agriculture in 1933, after a career spent mainly in agricultural colleges and at Edinburgh. There were three other Chairs associated with agriculture at the time, in Agricultural Botany, and in Agricultural Chemistry, the latter post being held by H. A. D. Neville, who was also Dean of the Faculty and a very powerful man, politically, in university affairs for a period of some 25 years. The third Chair was in Agricultural Economics, held by Edgar Thomas. In this capacity he was also responsible for the development of advisory work in economics and management and for the collection of agricultural statistics for the South Eastern Province. This was, of course, ten years before the formation of the National

Agricultural Advisory Service, and specialist advisory work was still centred on strategically placed university departments (e.g. Cambridge for the Eastern Counties, Newcastle for the North, or Nottingham for the Midlands). A Chair in Horticulture had also been created in 1932, and this department was developing rapidly at the time.

Robert Rae was a quietly spoken Scot, and one of the kindliest of men that one could ever wish to meet. He was a marvellous support and example to me in the early days when I was still finding my feet, and he turned out to be a true 'guide, philosopher and friend' on whom I could rely implicitly for help and advice. His career at Reading was relatively short, as it was interrupted by the war in 1939 when he was called up for responsible work in the Ministry before being sent out to Washington as Agricultural Attaché at the British Embassy. On his return, he was appointed to succeed Sir James Scott Watson as Head of the National Advisory Service, where he was duly, and deservedly, awarded his knighthood. But in his relatively short time at Reading, he played a large part in raising standards and modernising the structure of the courses – work which provided the base for the tremendous expansion which was to follow under the Professorship of Jem Sanders, who was appointed to the Chair in 1944 when Robert Rae resigned from it.

University Farm

One of Rae's first acts when he was appointed was to make out a strong case for a larger university farm, since the existing one at Shinfield was too small to provide the range and scale of enterprises needed for a growing department, and for the provision of research facilities which it was anticipated would increase considerably in the years ahead. Eventually a 380 acre site at Sonning, some three miles east of Reading, was purchased in 1935, the year before I appeared on the scene. It was bordered on the north west by the Thames, and on the south east by the main A4 London road. There were about 100 acres of alluvial, low lying land in permanent pasture along the river, while most of the remainder of the farm was on light, thin, gravelly soils, prone to acidity. It was in very poor condition, much of it in tumbled down grassland which had been farmed largely on a 'dog and stick' basis during the recession, though there were about 30 acres of so-called arable land along the London road, which were very heavily infested with couch grass, and every arable weed in the book. It was an excellent area for taking farm classes on soil management and improvement but I was very careful beforehand always to brush up on my weed identification before venturing out there with the students. This was before the days of systemic herbicides, and I well remember the long discussion in the farm office as to whether the whole area should be treated with sodium chlorate to kill the couch grass which would entail the risk of losing the following crop from the long-term residual effect of the chlorate. In the end, with the wisdom of Solomon, half the area was treated, and half was cleaned up as well as possible by continual cultivation and collection of the couch, which was taken off in cart loads. My recollection is that there was no residual effect on the treated half, but it certainly didn't kill off all the couch.

There was a large set of rather neglected buildings, quite well placed in the centre of the light land block, and the herd had been moved over from the old farm in the previous autumn, as had the sheep. There was also a large concrete tower silo, which had been put up sometime in the 1920s during one of the short-lived silage booms, which had periodically occurred almost since the turn of the century.

The farm really was in very poor condition, which from a teaching point of view was no bad thing, as the conditions were no worse than those which could be found on many other farms at the time. By a curious coincidence, I was to find myself responsible, some 15 years later, for taking over a very similar type of farm in an even worse state of neglect 30 miles further up the Thames above Oxford. The only real difference was that the major part of that farm above the flood plain was on very heavy Oxford clay instead of the light gravel at Sonning, which I think made it even more difficult to deal with, because of soil structure problems.

The dairy herd of Shorthorns had been brought over intact, but Robert Rae having previously worked in the Eastern Counties had contacts with Lord Rayleigh's Terling estate, which had the leading Friesian breeding herd at the time. He quite rightly saw the Friesian as the dairy breed of the future, and managed to negotiate for twenty Terling and Lavenham heifers to be provided to form the nucleus of a Friesian herd at Sonning. It was obviously a good advertisement for the breed which was then trying to establish itself as an alternative to the Shorthorns and Ayrshires which were still the dominant breeds. They fairly rapidly replaced the Shorthorns at Sonning, as was to happen in many other dairy herds over the next 15 years or so. As the heifers came in, the herd went up to some 70 cows, which was quite sizeable for a pre-war herd, and they were machine milked by two men and a boy.

The other main farm units were a sheep flock of some 250 half-bred ewes, and a flock of poultry which were in folds, containing about 30 birds each moved daily across short-term leys. There were about 40 folds, so it was a fairly labour intensive unit, with all food and water having to be taken to individual folds. But it was envisaged that the fertility left behind would be vital in helping to bring round some of the light land into a better condition for cereals. Folded poultry units were quite common on light land cereal growing farms in the south, having to some extent replaced close-folded sheep flocks. The low price of lamb and increased labour costs had killed off the old type of folded flock from about the time of the First World War onwards. Eggs were not much better, but they were just about economic when produced in this way, when the fertility left behind was taken into account.

Teaching

There were two main courses in agriculture at Reading when I took up my post in 1936: a full three year degree, and a two year diploma. A short certificate course of one year was being phased out. For crop and animal production, the degree and diploma students had the same lectures, and there were some 30 students altogether for lectures and farm classes. There were just four of us dealing with mainstream agriculture: Robert Rae, David Black, Kenneth Campbell who covered the animal production aspects, and myself. Then there were subsidiary lecturers in machinery, agricultural botany, chemistry, surveying, economics, and so on.

It had been agreed that I would take over from David Black all the basic non-animal topics, such as soil management, fertilisers, soil improvement, cultivations, and also grassland, leaving him to deal simply with the production of the main crops.

As soon as I arrived about three weeks before the start of term, it became a race to get out a set of lecture notes in time. They had to be pretty full notes, too, since I certainly hadn't any experience of speaking off the cuff, and didn't want to run the risk of being caught short, or drying up at a critical moment.

Fortunately I was always able to keep about one week ahead in that first term in preparing lectures, but it was quite a struggle. I had kept very full notes of Jem Sanders' lectures at Cambridge, and these proved to be invaluable, as he was probably the best teacher in agriculture that there has ever been – certainly in that era, at any rate.

It wasn't only lectures that had to be prepared for, but also practical farm classes at Sonning once a week. If anything these were more difficult than lectures, for one had to be prepared for any sort of question in a practical class, not least, tricky ones from farmers' sons trying to prove to the rest of the class that they knew a good deal more than I did, which was sometimes the case in the early days.

At times, during the preparation period, I wondered why on earth I had ever applied for the job, and got myself into this frightening situation, and thought how much easier everything would have been if I had accepted George's offer to take on the farm in Cornwall. Of course, it wouldn't have been, for there would have been equally frightening situations to have been faced up to, or possibly even worse ones. In the event, once term had started, and things were under way, it wasn't too bad, though there were sticky moments at times with some of the more boisterous elements in the student body.

I think that any teacher would confirm that annual intakes of students tend to fit into three categories – good, poor, or plain indifferent. Every now and then, there comes a vintage year, with an excellent group of students, often with one or two outstanding personalities setting the pace, and raising the level of the rest. Alternatively, it can be the opposite, with one or two leaders being anti this or that, and taking the weaker ones along with them so that the whole tone of the course suffers adversely. And then sometimes, it is a year without any real personalities in it at all, and it becomes difficult to strike any sparks of enthusiasm from a rather dull and mediocre group, and at the end of the year, one breathes a sigh of relief, and hopes for better things next time.

I was, on the whole, lucky that I ran into a real vintage group in my first year. I say on the whole, because being an outstanding group, they were also a pretty exuberant lot, and always prepared to take the mickey out of an inexperienced young lecturer if he gave them the chance. But against that, I was able to learn a great deal from them, particularly in the field of human relations. Their exuberance was an expression of high spirits, and was never at all malicious. Their leader was Ken Blaxter – the late Sir Kenneth Blaxter – probably the most outstanding animal scientist of his generation, and with quite the most brilliant mind of any student that I ever taught. It was almost impossible not to give him 100 per cent marks on any exam papers that he took – every fact was there, set out in perfect order, and in impeccable English.

He had round him a very lively group, several of whom were farmers' sons whose fathers had successfully survived through the depression, and they were, practically, a very knowledgeable group. They used to sit in the back row of the lecture room, ready to pounce on any statement from the lectern with which one of them might disagree. It wasn't the sheer bloodyminded barracking that I had had to sit through in chemistry lectures at Cambridge, which often ended in a near riot, but a bit of good-natured light relief brought in to relieve the tedium of sitting at the receiving end of an hour's lecture. If it looked like getting to the point where it was interfering with the concentration of the other students, they were just told to shut up and keep quiet, which had the desired effect when the ultimatum was delivered in one's best parade ground manner. Outside the classroom, farm classes enabled much closer contacts to be made, and even friendships formed which have lasted over the years.

The group's most audacious exploit followed a farm class one summer afternoon. The class had ended, and I left them in the yard to disperse while I went back to the farm office for about half an hour. Returning to the yard to drive back to Reading, what should I see but my dear little box-like 1928 Austin 7 (cost £7 second hand in 1937) reposing two tiers up on a pile of straw bales onto which the group had man-handled it. It took about half an hour for the farm staff and myself to get it down again. After a few moments of fury and a feeling of anger at being made to look ridiculous in front of the farm staff, I was fortunately able to see the funny side of it and to appreciate it as being really rather a good practical joke – and quite an outstanding physical achievement at that. In the lecture next day, I congratulated them on their physical prowess, but said quite firmly that in future they were not to play fast and loose with my personal property, and they never did again.

Experimental Work

One of the conditions of my appointment was that I should combine some research with teaching, and I decided to follow up my Cambridge work on the manuring of permanent pasture, on the low lying riverside pasture on the Sonning farm. So in the spring of 1937, I laid down quite a large scale trial comparing different levels of nitrogen, phosphate and potash under a frequent cutting regime. It was designed as a long term trial, and it confirmed during the two years that I managed it that, as at Cambridge, nitrogen was much the most important element in increasing yield, even with the small amounts then being used.

I also became closely involved with some work on the use of pneumatic tyres on farm vehicles, which the Rubber Growers Association under its Director, Sandy Hay, was financing at Sonning. It is difficult to believe now that at the time virtually all tractors were still on spade lugs, and farm carts and wagons were on iron-rimmed wheels. Pneumatic tyres were certainly coming in for tractors, and the Oxford Institute was conducting research in this field. Some farmers, too, were experimenting with old car axles on farm carts, but no figures were available to prove that pneumatic tyres were more efficient in terms of reduced draft.

To obtain these, J. B. Passmore and Sandy Hay had lengthened the shafts of a conventional dung cart to enable a dynamometer to be inserted between the horse and the cart, the tractive effort being applied through a pair of plough chains and a whippletree. It was all a bit elementary but in practice it worked surprisingly well. As the junior member of the team, it was my job to walk alongside recording the dynamometer readings over the given distance of 200 yards, which wasn't very easy, because of its position between the shafts. Three different types of cart were used, loaded with 17.5 cwt of basic slag and tested over a grass field, and arable land after a mangold crop. Conventional iron wheels were compared with ordinary pneumatics, and also with pneumatics fitted with roller bearings.

There was little difference between the two types of pneumatic, but when it came to the iron wheels, there was a dramatic reduction in the drawbar pull attributable to the pneumatic tyres. This came down from 275 lb to 150 lb on the grass, and from 475 lb to 350 lb on the arable land. It was very noticeable how much more the iron wheels cut into the arable. It was a little surprising that the difference in the readings was not greater still. Two conclusions were drawn: firstly, pneumatics were much more efficient on farm carts, and secondly, that when they were used, load weights might be increased.

Several other possible uses for small pneumatic tyres on

farm equipment, such as sack barrows, water carriers, and poultry folds were also investigated. The farm poultry folds had to be moved onto fresh ground daily, and a set of portable wheels were designed which would fit under the cross member. A simple levering device lifted the fold which could then be pushed by hand onto new ground. Poultry folding soon fell into disuse, so unfortunately the system never caught on.

Reading was a good place to be based from the point of view of keeping abreast of developments in the agricultural world. For one thing it was the regional centre for specialist advice, with Wye College as a subcentre, and regional conferences of advisory officers were held from time to time which we attended. I well remember the excitement caused at one of these by the first announcements of the results of feeding iodinated casein to dairy cows which showed significant increases in yield under some conditions. Even then doubts were expressed about the long-term effects of stimulating natural hormonal systems to work at levels higher than nature had intended – a scenario that was to be repeated in more exaggerated form by the introduction of BST some 40 years later. In the event, iodinated casein never came into general farm use because of the difficulty of controlling intake, and the variable response of dairy cows to it, and the risk to the health of the cow, which appeared to be considerable in the light of several reported deaths. It is strange that this work seemed to have been forgotten when the BST controversy was at its height.

There was also the National Institute for Research in Dairying at Shinfield, loosely attached to the university, with a great deal of interesting research in progress, both on fundamental issues such as clean milk production, and on applied research in the field. Up the road at Oxford, the first of the Conferences on Mechanisation had been held in January 1936, with a second one in 1937, and though I did not personally go to any of them, quite a lot of information of progress in this field filtered through, and we visited the St John's College farm at Long Wittenham, where the School of Rural Economy and the Institute of Agricultural Engineering were looking into the techniques and economics of mechanised farming without livestock.

In addition to this, the students' Agricultural Club was very active with weekly meetings, at which most of the leading agricultural scientists, and pioneering farmers of the day were invited to speak.

Reading, too, was the leading light in the Agricultural Education Association, which was one of the first of the specialist professional agricultural bodies. For many years its annual conferences provided a wonderful opportunity for educationists, scientists and advisory officers to get together and exchange information, and to keep themselves up to date on developments right across the board. Sadly with the later growth of more narrowly based professional societies of a specialist kind, such as the British Grassland Society and the Society of Animal Production, the AEA lost much of its influence, and its future as an Advisory Service body now hangs in the balance. But at the time, it was in its heyday, and I found its meetings both an excellent opportunity for getting to know people since it was a very friendly organisation, and also invaluable for learning about many different aspects of the industry.

I was still very concerned that I had done virtually nothing with machinery and that I really ought to try to get rather more experience with tractors and their operation. In the autumn of 1936, the Ford Motor Company advertised an intensive week's training course just before Christmas on tractors and ploughing which the Department paid for me to go on.

Unfortunately the land was frozen solid so there was no opportunity for ploughing but the course was good on mechanics and I learned a lot about the internal workings of the tractor. Of the dozen or so attending the majority were farm workers sent by farmers changing over from horses to tractors.

I was given an opportunity for further extramural activity that year when, in the early summer, Robert Rae received a letter from Cadbury's at Bournville asking if he could recommend to them a young agriculturist to accompany parties of their young employees for three separate weekly trips on a converted barge on the Shropshire Union Canal between Wolverhampton and Chester. Cadbury's (like Rowntree's) was founded by a family with strong Quaker connections and had always been model employers with a strong commitment to the wellbeing of their workforce – hence, the building of Bournville. They were also very concerned with the general education of their young employees, and catered for this through evening classes, a youth club, sporting activities, and so on. They had converted an old canal long boat by covering it over, fitting it out with tables and benches, a galley and storage area, and this was used as the base for weekly trips on canals for the young employees. They were given the week off work and the trips were essentially an educational exercise. It was a most imaginative innovation in extramural education for its time.

My task was to talk to the boys on farming and agriculture in general as the boat meandered its way through the countryside, pulled by a docile Shire gelding, who always seemed somewhat bored by the proceedings. There was a small payment offered, which was an additional attraction, and I thought that it could be good experience, as I did not know anything about farming in that area.

We slept out in tents at selected sites on the way, and had our meals on the barge. There were about 15 boys on each trip, and in charge was a Youth Leader from one of the clubs at Bournville. But the real key to the success of the whole enterprise was a marvellous character called called Ernie Nutt, who masterminded the logistics, did the cooking, oversaw the erection of the tents and had the boys completely under control in a quiet authoritative way. He had been a Sergeant Major in the First World War, and was one of those fortunate people who can put their hand to anything, and had a tremendous breadth of knowledge on all sorts of subjects, and a great deal of experience of the world in general. He stood no nonsense from anyone, but the boys worshipped him. Naturally as a complete stranger, none of them quite knew what to make of me to start with, but Ernie and I seemed to get on well from the word 'go', and it wasn't long before each party of boys ceased to regard me as some sort of schoolmaster sent to give them lessons and accepted me as someone on their own level.

The day's routine was that, after breakfast there would be a talk and discussion on some topic or other, then a break, and another discussion. The afternoons and evening were occupied by visits to a variety of local centres of interest. We went, for example, round the railway workshops in Crewe, down a salt mine at Northwich, visited Harper Adams College (which interested me as I hadn't seen it before), had a conducted tour round Chester, and climbed to the top of Beeston Hill for a geography lesson. It was an imaginative programme, and I learned a great deal from it myself, and it was invaluable for the boys.

On the whole, the weather behaved itself, and it was really quite an idyllic experience to be drawn steadily and noiselessly through the Staffordshire and Cheshire meadows on a sunny July morning. Every now and again, it would be one's turn to lead the horse along the towpath, which provided an opportunity for quiet contemplation of the countryside. At that time, of course, the area of arable land in a county like Cheshire was extremely small, and the land along the canal was practically all under permanent pasture. Some of it was well farmed, with tidy hedges and closely grazed swards reasonably free of visible weeds. But some was infested with docks and thistles, with yards-wide hedges encroaching into the fields and small scrub infiltrating here and there. There were no grassland herbicides available then, and the only control for docks and thistles was by cutting at fairly frequent intervals. The time to cut thistles for maximum effect was just after they came into flower, when the root reserves were at their lowest level in the growth cycle. When fields were being cut for hay, there was generally insufficient growth, because of low fertility levels, to cut much before mid-July and that was often a bit too late to control thistles very effectively, as some seed would already have been shed. It was permanent pasture like this for which the pre-war grants for basic slag and lime application were designed, in order to try to bring them to better levels of fertility, in case they were to be needed for higher production in the event of war.

I greatly enjoyed those three weeks, and learned much about how to get on with people coming from a very different background from my own.

It is all too easy when looking back into the remoter areas of one's past to remember the good things, and to forget or pass over the less pleasant ones. Summers in retrospect were always full of sunshine, and life seemed to be lived at a more leisurely pace, with plenty of time for recreation and for sport. Even allowing for such rose-tinted impressions, however, I do look back on my two years at Reading with nostalgia and pleasure. After the first few rather hectic and traumatic weeks of scrambling to get on top of lecturing and taking field classes, and of getting to know the students as individuals and (one hopes) of getting accepted by them, my previous worries about being able to cope with the job were largely forgotten. The move into Wantage Hall at the end of the first term was a great help, as I was immediately placed in close contact with student life on a day to day basis, and with the other staff members who lived there. Apart from the warden these were the rather formidable Professor of Agricultural Chemistry, H. A. D. Neville, with whom I always found it rather difficult to establish a rapport, and Jock Thompson, a fairly junior lecturer in agricultural botany, who was three or four years my senior. We became close friends and spent a good deal of time together particularly in the evenings when, in company with another youngish lecturer in classics and a teacher from Reading School, we would visit various hostelries in the neighbouring countryside for a game of darts or bar billiards. J. B. Passmore laid down the maxim that staff should not drink in the same pubs as the students – probably a good one – so, as we had cars, we went outside the town, leaving the town pubs to them.

In addition to evening excursions, there were university social events of various kinds. I well remember a summer dance in one of the women's halls which was unfortunately a soft drinks only occasion. (Presumably the warden could not trust her girls to hold their drink well enough to protect themselves from the amorous attentions of the male students.) About halfway through the evening Jock and I and one or two men students were hustled off very surreptitiously to the bedroom of one of the girls, a most extrovert Swedish student who obviously knew her way about the world pretty well already. Safe from the eagle eye of the warden, she produced a couple of bottles of whisky hidden beneath some intimate articles of clothing in her chest of drawers. Glasses were rustled up from neighbouring rooms, and the once dry dance was quickly well lubricated. Jock and I decided afterwards that J.B.'s dictum about drinking in pubs with students did not really apply to girls' bedrooms in halls of residence!

My principal recreation over those two years, however, was golf, and I was fortunate to be able to get into the Sonning Club, just across the London Road from the university farm (no waiting lists for club membership in those days, thank goodness).

There were also various formal occasions when Wantage Hall was used to entertain distinguished visitors, sometimes for dinner either before or after addressing some student society or other, and this was an opportunity to meet, or to listen to quite a number of very interesting people. One such, who made an impression of quite the wrong kind, was Professor J. B. S. Haldane, who had, I think, a Chair at the London School of Economics at the time. He turned out to be not only arrogant, but also rude, spending the entire meal denigrating redbrick universities in an intellectually patronising way. J. B. Passmore, who was acting as host, was certainly not one to suffer fools or pretentious academics gladly, and one of the more entertaining aspects of the meal for us as onlookers, was to watch him trying to contain himself in a manner befitting a host. This, to his great credit, he managed to do, but his language after his guest's departure was quite the most lurid that Jock and I had ever heard – and that is saying something, for his vocabulary was notorious throughout the university.

Another distinguished visitor, though this time at the

luncheon for the recipients of honorary degrees, was Walter Sickert RA. He was already at quite an advanced age, and was well known for unconventionality, which he appeared determined to live up to, and it was clear that he did not intend to be impressed by the occasion. Presumably he suffered from foot problems for he turned up wearing some kind of carpet slippers and dressed in appropriately dishevelled clothes – appropriate for a highly distinguished artist, that is. He was supported by his wife, who seemed rather concerned as to what he might get up to next. He spent a considerable time over lunch crooning quietly to himself, but no doubt greatly to the relief of the university authorities, he behaved impeccably during the degree ceremony itself.

A still more distinguished visitor the previous year had been the Prime Minister, Neville Chamberlain, who was also receiving an honorary degree. He had been approached about accepting the Chancellorship of the university, but had declined owing to pressure of work, but had suggested Sir Samuel Hoare, the Foreign Secretary, instead, and he was duly elected. He turned out to be a very conscientious Chancellor, I believe, though he fell from power within two years as a result of the notorious Hoare/Laval Pact, which would virtually have recognised Mussolini's annexation of Abyssinia.

I have always felt, incidentally, that history has not been altogether fair to Chamberlain over his role in the Munich crisis. I don't believe for a moment that he was as gullible as he has been made out to have been (he appeared to be quite a shrewd old bird when viewed at close quarters, in spite of his high collar and old-fashioned appearance). It was obvious that he had to play for time at Munich, since the country was still totally unprepared for war in 1938, and popular feeling was still very much against it – and Hitler certainly knew that as well. The pacifist policies of the socialist Government in the late '20s, followed by the economic crises of 1929 and 1931 and their aftermath had led to severe reductions in the strength of the armed forces, and to a failure to modernise their equipment.

There was still a very strong antipathy in many quarters to militarism, the result of a very understandable revulsion at the horrors and the senseless loss of life in the First World War which was still very fresh in people's minds. Rearmament did not really start to get under way until the mid-1930s when Hitler had begun to show his true colours, and it takes a considerable time and a great deal of money to build up a weapons and aircraft industry capable of delivering sufficient equipment to three different services to enable them to participate successfully in a war. Germany and Italy had several years' start over Britain, and had had an excellent opportunity to test out their weaponry in the Spanish Civil War in 1936.

It is true that Chamberlain appeared to go over the top, so to speak, in his 'Peace in our time' speech on his return from Munich, but the fact that the Government continued to press ahead with rearmament as fast as it possibly could, suggests that he did not really believe that the agreement could stick for very long. The public reaction to Munich showed that, at the time, he did have massive public support for his attempts to hold the peace.

The years from 1935–38 had been a period of intense political activity, and the Spanish War, the occupation of Abyssinia, the persecution of the Jews in Germany and Hitler's sabre-rattling had finally shown the old League of Nations to be a completely ineffective body for controlling aggression and maintaining peace in the world. Whilst another war on a global scale less than 20 years after the dreadful destruction and slaughter of 1914–18 seemed almost inconceivable, there appeared to be no other way by which the two dictators could be restrained. The majority of people still hoped that diplomacy in some way could win the day, and that even if one or two countries were lost, the appetite of the dictators for territorial expansion would be satisfied. Optimists believed that, because the German and Italian regimes *were* so repressive, it could not be long before the dictators were overthrown by internal coups, to be replaced by more democratic governments. I well remember the long discussions, both as an undergraduate at Cambridge, and then later in the Senior Common Room at Reading as to what might be done, and the sense of frustration that, as individuals, there was nothing we could effectively do about it. So, in practice, one got on with one's job, enjoyed life as much as one could, and tried to exclude the world's problems from one's thoughts as much as possible. This was the case right up to the Munich crisis, though by that time my life had taken another turn, and I was on a ship in the middle of the Mediterranean when Chamberlain was waving his piece of paper to the crowds at home.

As far as agriculture was concerned, the international uncertainty was to some extent beneficial, because it turned people's minds back to the food shortages of 1917, especially as it was known that Hitler was basing his naval expansion on fast surface cruisers and a large U-boat fleet. Though no actual direct measures to increase food production were taken at the time, legislation was passed which aimed at improving the fertility of the land through grants for the application of lime and phosphate, and for drainage and the ploughing of old grassland. This had quite a marked effect on improving the confidence of more progressive farmers.

At the same time, in 1937, a special department was set up in the Ministry to formulate plans for an immediate all-out food production effort which could be put into effect rapidly if another war did come. This plan was based largely on the experience of the 1917–18 campaign initiated by Lloyd George to meet the food crisis of those years. It proved of inestimable value when war came in 1939.

On a personal basis, the early summer of 1938 saw another change in direction brought about by a letter that Robert Rae had received from the Sudan Government Office in London about the establishment of a new agricultural faculty in Khartoum University. It was to be located at Shambat, just outside the city, with new buildings on a university farm, and they were looking for a youngish man to set it up and become Head of the Department. Robert asked if I was interested, but I felt that it was not quite my type of job. Jock Thompson, however, showed a lot of interest when I told him about it, and he applied and was appointed to the post.

Before I decided it was not my cup of tea, I had made some enquiries about the Sudan, and found that there were vacancies for two Inspectors of Agriculture to join the field service. This consisted of some 40 British staff, whose jobs were mainly supervising the production of either cotton, sorghum, or, in the South, subsistence crops. It was a small, select service, and the jobs appeared to provide a lot of freedom and independence in a vast country which was then developing quite fast economically.

I was attracted to it for three reasons: firstly, the pay and allowances were far higher than any job which would have been available at home; secondly, there was three months home leave every year after an initial two year probationary tour; and thirdly, my three year appointment at Reading would be up the following year. The employment situation had not improved at all, so if I was not reappointed, I might find myself out of a job. But really at the back of my mind was the thought that within about seven years, I would have been able to save enough money to come home and buy a farm – I had by then decided that I did really want to farm, but only if I could do it with my own money. So in rather a hurry, since the application date was very close, I filled in the form, and almost before I knew where I was, I had been summoned for interview and was offered one of the jobs. This I accepted, with a departure date set for the middle of September.

Looking back on it now, I am rather surprised by two things. Firstly, why I should have been prepared to make such a drastic move after only two of my years at Reading and from a job that I was enjoying so much, and to which I might reasonably have expected to be reappointed when the time came. Secondly, why I did not appear to take the possibility of another war into consideration, with its obvious risks of being cut off from home completely for a long time. I suppose the answer to the first question was 'itchy feet', money, and a desire to see more of the outside world, while at the same time, being able to spend three months at home each year. One has to bear in mind that the opportunities for moving around the world for someone with no money were more or less non-existent in those days, and the salary was a definite incentive. With regard to a possible war, I suppose it well illustrates the attitude of my generation at the time. We were worried and perturbed by the international situation, and by our apparent inability to do anything about it, but we never, even in the summer of 1938, *really* believed that it would come to war, certainly not until the Munich crisis.

FOREIGN INTERLUDE

Sudanese Experience

With many feelings of regret, not least at parting from Robert Rae who had proved to be a true friend, and my other colleagues who had been so helpful in getting me over the first hurdles of a working life, I left Reading early in August. I then had a month's holiday before sailing from Birkenhead for Port Sudan on 16 September. This was the same date on which, as a timid boy 11 years before, I had arrived for the first day at my public school, and I think that my feelings were much the same in each case. There was a sense of suddenly being cast adrift from a previously quiet and settled existence into a new world of uncertainty in which the future could not be foreseen. On the train to Birkenhead, I remember reproaching myself for being such a fool as to have left Reading so soon without even enquiring into a possible future there.

The *Salween* was quite a small boat carrying a lot of cargo and some 40 passengers, bound for destinations from Marseilles to Rangoon. There were several British wives and young children going out to join their husbands for the winter. In general wives did not stay out during very hot summer months, though a number of European women, such as nurses, were there throughout the year.

Much of the day was spent lying in the sun on deck desperately wrestling with Arabic textbooks, since one's promotion and pay prospects depended on passing an exam in the language before the end of the two year probationary period. Having no gift for languages, I found it very difficult, but was helped quite a lot by one of the wives, who had been in the country for a fair time, and had picked up a reasonable vocabulary, even though it was mainly household Arabic. But that turned out to be useful when I found myslf having to deal with no less than three servants, two of whom had no English, within a few days of arrival.

We were in the middle of the Mediterranean, somewhere near Sicily, when the Munich crisis blew up, and for the first time I began to wonder quite seriously what would happen if war was actually declared. Some of the ladies were a bit alarmed that if it did, the Italians might suddenly sally forth and send us to the bottom.

There was, therefore, very great relief at the outcome of Munich, which came before we reached Egypt, and a feeling that the Prime Minister had done a good job in averting the crisis, even if it might be only in the short term.

Port Sudan was hot, very hot, when we finally got there after 16 days – a high humidity Red Sea heat which left one wet with sweat and a feeling of exhaustion.

The Sudan Government at the time was a so-called condominium between Britain and Egypt, whereby the country was theoretically jointly controlled by both, with equal opportunities for employment for both nationalities. In practice, nearly all the senior administrative posts, and most of those in the specialist departments, were held by Britons, on the basis that there were no Egyptians with adequate qualifications for the higher jobs. This was probably true so far as the administrative service was concerned, though perhaps less so for the service departments, such as Public Works, Railways and so on, where most of the second tier jobs were occupied by Egyptians or Sudanese. The administrative service, which really ran the country through the District Commissioners, Regional Governors and the central headquarters in Khartoum was staffed almost entirely by Oxford or Cambridge 'blues' of one sort or another. Hence the old jest that the Sudan was, 'a country of six million blacks ruled by a handful of blues' – and it was only a handful, too, for the total number of British nationals probably did not exceed 300 or so, of whom about 40 were in the agricultural service.

The Sudan, for those who know little of the country, covers a vast area, stretching from the Egyptian border in the north some 1,300 miles south to the border of Uganda, and nearly 1,000 miles east to west from the Red Sea to Chad. Most of the North is desert apart from the Nile valley, and then there is a large belt of savannah-like country in the centre, merging into equatorial higher-rainfall areas in the South. Khartoum lies at the confluence of the White and Blue Niles, some 500 miles south of Egypt, and about the same distance west of the Red Sea. The White Nile originates in Lake Victoria in Uganda, 1,000 miles away, and is a slow-flowing, sluggish river, which gets almost lost in the vast swamp area known as the Sudd, halfway up to Khartoum. This is virtually impenetrable and it forms the dividing line between the Arab North, and the Negroid South of the country. The Blue Nile rises in the mountains of Ethiopia in the south-east, and is a fast-flowing stream with big seasonal variations in its flow rate. The Nile water is absolutely vital to the economies of both the Sudan and Egypt, so that a number of dams have been built over the years to regulate its flow, particularly on the Blue Nile. These really serve three purposes: firstly, to prevent waste of water, by holding back the Blue Nile flood, and then gradually releasing it over the year; secondly, to provide irrigation for the production of cotton, on which the whole of the economy of the country depends; and thirdly, for the generation of hydroelectric power.

The largest of the dams was built at Sennar, some 200 miles south of Khartoum and this came into full operation in 1925, its construction having been delayed by the First World War. This brought under cultivation an area 100 miles in length and 35 miles in width, with over 800,000 acres irrigated. More areas have been brought in since then, and other smaller dams built. The rainfall over the country varies from 10 inches a year in the North to 20 inches in the South, but it all falls in 3 months, and in some parts of the North, there may be virtually none at all. Cotton is by far the most important crop, followed by sorghum and maize, and the South used to export considerable quantities of gum arabic and sorghum.

But due to the civil war between the North and South, which broke out 25 years ago, the country is now chronically short of food. The war has been caused primarily by attempts of fanatical Arab Fundamentalist elements in Khartoum to impose Sharia Law and the Arabic language, religion and culture on the Negroid tribes of the South, many of whom are Christians. Originally this was resisted strongly by the People's Liberation Army (PLA) which was formed in the South, and resulted in a virtual stalemate.

Now, regrettably, the Southern forces have been split by tribal rifts, so that there is less united effort against the North, and even greater disruption and starvation in the southern areas. It is impossible to believe that the North can ever be successful in imposing a completely alien culture on the Negro peoples, and the tragedy of the whole mad affair is that the North is spending huge sums on its armed forces, money which should be spent developing the country to a point where it can both industrialise, and also develop its agriculture to feed its growing population.

To one who remembers this huge and largely peaceful country where the services functioned efficiently, the people were happy and well fed, hospitals and schools were quite well equipped and improving year by year, and agriculture earned the foreign exchange required to run the economy, the situation today is enough to make a strong man weep. It is all the more tragic in that the Sudanese are a noble, proud, and most attractive people, whether they come from the Arabic peoples of the North, or from the Negroid tribes in the South, and it is dreadful to contemplate the level of racial and religious hatred that has built up since Britain and Egypt relinquished control in 1953. Even if peace can be restored, it will take many years for the scars to heal, and the future may well lie in the division of the two parts of the country into separate states with the South linking up with Uganda in some way.

It is fashionable nowadays to scoff at the role played by the

British – or the French, Belgians or Portuguese – in the development of Africa in the 19th and early 20th centuries but it worked, and provided peace, food for all, and a degree of prosperity to the countries concerned, even if it did mean in some cases the exploitation of their resources. Sudan, Angola, the Congo, and even Uganda are sorry examples of what can happen when evolution is replaced by revolution.

The train for Khartoum left in the evening of the day we arrived, and the first class coaches which we were privileged to occupy were air conditioned to some extent. At that time most of the economy of the country depended on the railways, and particularly on the Khartoum/Port Sudan link, since that was, and still is, the only port for the export of cotton to the outside world, and for the import of food and raw materials. There was no lorry transport, and the use of the Nile as a medium for transport was restricted by the cataracts to the North, and the barrier of the Sudd to the South.

The railways were run most efficiently by a small number of British and Egyptian engineers, and a very well-trained Sudanese labour force. The trains, of course, ran very slowly along the single line track, and a constant look out had to be kept for sand blowing across the line, especially on the Khartoum link, which runs for some 500 miles through largely desert country. We left in the evening and ambled up through the Red Sea Hills, and then on across the huge sandy plains, at, I would guess some 15–20 mph reaching Khartoum in the afternoon of the second day.

Two days of indoctrination followed at the HQ of the Agricultural Department. All senior officials visiting the capital stayed at the rather majestic Grand Hotel on the banks of the Blue Nile, not far from the official residence of the Governor, where some 50 years previously, General Gordon had met his death at the hands of the Mahdi's followers. There was a romantic air about the place, with its palm trees, acacias and beautifully maintained gardens, and masses of servants at hand whenever one wanted anything. It felt like Somerset Maugham country, as, in a way, I suppose it was.

First Posting

I was posted to the Gezira Research Station at Wad Medani, about 120 miles south of Khartoum, as the Junior Inspector of Agriculture. Medani was a large town on the Blue Nile, and the administrative headquarters of the Blue Nile Province, lying roughly at the centre of the Gezira – the huge area of flat land between the two rivers as they converge to the apex of the triangle of Khartoum. With the opening of the Sennar dam, nearly a million acres of this land had come under irrigation, and Wad Medani had become the commercial centre servicing the area and had grown rapidly in importance and size.

The Research Farm lay about two miles out of the town, and comprised 800 acres, all irrigated. There were five specialist research departments, each run by a senior British scientist, with usually a British assistant. The agricultural department ran the farm, the dairy herd and a sheep flock, with a Senior Inspector of Agriculture from the field service, and a Junior Inspector, my job. The Station was virtually a community on its own, with very well-equipped laboratories, bungalows for all the senior staff, and additional housing for the Sudanese technicians and administrators. The bungalows were well-equipped with large gardens, staff quarters, and a couple of stables for horses.

The oldest and least comfortable of the bungalows was allotted to the Junior Inspector and, being unmarried, he was expected also to provide accommodation for one of the young scientists. So I suddenly found myself in possession of quite a large house, three servants – a houseboy, a cook and a groom – and within a week, a lodger, and a third person to feed. The lodger was a young plant physiologist of about my own age, who was very easy going and a pleasant person to be with, and the other one who ate with us, but slept elsewhere, was a young entomologist, who was also extremely easy to get on with. It was a good arrangement as it provided company and allowed us all to live economically. The groom was required as I was expected to use a horse for going round the farm in the mornings to supervise work in progress. So within a week I had also acquired a horse, of which there is more to be said later.

My senior, Edward John, was just over 40, with some 20 years' experience behind him, much of it spent in another cotton growing area in the East, towards the Eritrean frontier. He was rather highly strung, and not too fit, and tended to operate on a rather short fuse at times, but we got on well together, and I was never, fortunately, on the receiving end of one of his minor explosions. He was very sound on his practical farming in relation to the crops that were grown on the farm – cotton, sorghum, alfalfa, and a type of haricot bean (Lubia). He was also very keen on developing the dairy herd of some 40 cows, which was kept primarily to provide milk for the staff at the Station. He was in the process of trying to build up a high- or relatively high-yielding herd of the local Zebu type cattle, by buying in what appeared to be promising cows from other small herds kept by the cultivators on the Gezira Scheme. He had collected quite a good nucleus of cows over the past three years to serve as a foundation, but the problem was to find suitable bulls as, naturally, there was no recording system, except on the farm.

Even recording on the farm was difficult, because of the native belief that a cow would never let her milk down unless she had been first given a suckle by her calf. Unfortunately, the cows had become conditioned to this from being milked that way as heifers and Edward had not solved the problem, because the stockmen would always try to sneak the calf back again when one's back was turned. So the recorded figures were unreliable, as one never knew just how much milk the calf had taken. Nevertheless, there were some very useful cows in the herd, and we achieved our first 1,000 gallon cow when I was there. Edward John took the view, quite rightly I think, that there was no point in trying to take a short cut by importing a high-yielding European breed – with all the risks that that entailed of lack of heat tolerance and extreme susceptibility to disease – to use on completely unselected cows. His intention was, firstly, to see what the cows were capable of doing, given the best management and feeding, and then, secondly, to build up a herd of reasonable uniformity, using one or two home bred bulls from the best cows. Regrettably, I never heard how the project progressed, but I suspect that it did not get far after his departure.

Apart from Edward John, there was the Sudanese farm manager, Awam Nimr, one of the finest people that I have ever had the pleasure of working with. He came from a landed family in the North, and had been trained in Egypt, but had got into some sort of trouble in the early '20s, when there was unrest in favour of Egypt, mostly at student level. This had given him a black mark, and up till then, had probably prevented him from being promoted to a higher position in the agricultural department, to which his natural ability should have entitled him. I am glad to say that, as the country moved towards Sudanisation, he did get promotion, and became Head of the Gezira Cotton Board after the British withdrew in 1953, and then, I believe, Director of Agriculture.

At the time, I found him tremendously valuable in helping me with the technicalities of tropical crop production, and the management of the land under irrigation. He was also extremely useful in providing advice with regard to handling servants, and local customs. But his true value really became apparent when, only six weeks after my arrival, Edward John went down with acute appendicitis, from which he nearly died, and was sent back to England for six months to recuperate.

So I found myself in charge of running an 800 acre research farm for a group of high-powered scientists, with only some six weeks' experience of tropical agriculture. It was then that I appreciated, perhaps for the first time, the benefit of the Cambridge apprenticeship. The training that I had received in applying principles to practice and in weighing up the merits of different alternatives before coming to a reasoned decision really began to bear fruit. I realised that the principles of plant growth and crop management were exactly the same for cotton or sorghum as they were for wheat, and the basis of feeding the cows was the same for Zebus as for Shorthorns, even if the ingredients were cotton cake and millet, rather than ground nut cake and cereals, and

I had Awam Nimr at hand to guide me along if I needed help. He was the perfect Staff Officer, never for a moment giving the impression that he was, in some way, the person in charge, but always building me up with the senior farm staff as the real boss.

In fact, I think that we both thoroughly enjoyed the six months that we had together running the farm; he, because he probably felt that he was having a good deal more say in the management than he would have with an older British official in charge; and me, because I had complete faith in his technical and managerial efficiency which prevented me from dropping any bricks from inexperience, and also because I was relishing the experience of running quite a sizeable show of my own for the first time in my life.

I had, I like to think, quite quickly established a good rapport with the heads of the research departments, and with the Director of the Station, who was a plant breeder, and was very helpful in every way, not trying to interfere as he might so easily have done under the circumstances. Our job on the farm was really two fold: primarily to provide research facilities for any experiment that the scientists wished to carry out – and there were a lot of these that required careful management in regard to land preparation, sowing, crop treatment and harvesting. Our other function was to run the remainder of the farm as efficiently as possible, whether it was growing cash or forage crops, or managing the dairy herd or flock of some 200 ewes on alfalfa leys.

One important principle was not to allow the scientists to make sudden and unreasonable demands for the services of the farm staff. They had to let us know in good time when they were likely to want particular operations carried out, so that we could plan our own programme of work in advance. Awam had two or three well-trained technicians under him, who were responsible for actually laying out the plots and carrying out the field operations. Otherwise, there were few regular farm employees apart from the stockmen and two mechanics to look after the equipment. For hand labour, there was always a pool of casuals who appeared from nowhere at 6.30 each morning outside the farm office in the hope that there might be a job available that day. If there was no work, they disappeared as quickly as they had materialised.

The routine for the day was a 6.30 a.m. start at the farm office, when men were taken on and put to work on a programme which we usually settled the previous day – one of the advantages of having a climate in which you could rely on the sun shining for at least nine months on end. Then it would be a visit to the dairy, and possibly the sheep and to any jobs which were starting up close to the buildings. Then it was back to a good breakfast at 8.30 to 9.30, after which it would generally be a ride round the farm to check on irrigation, and the other jobs in hand. This was followed by some time in the office discussing the next day's work or policy matters with the farm manager. The real day's work ended at 2.00, when it was getting extremely hot, and it was time for lunch and the afternoon siesta till 4.00 when it was beginning to cool down again outside. Temperatures did vary according to the time of the year, but would generally be in the range of 30–40°C, and occasionally slightly higher; but with a low relative humidity, the heat was reasonably bearable. One could play squash, for example, in the open court at the club in Wad Medani at a temperature of 35°C or more without undue overheating.

After 4.00 there would be some three hours of daylight for recreation, which consisted of golf, tennis, squash, riding, or even polo, if you were that way inclined, which I was not. The sun would set at 6.30–7.00, often with glorious sunsets, and then it would be time for a bath, a drink, dinner at 8.00, and an hour or so on Arabic, before bed at 10.30. All the houses were netted against mosquitoes, and in addition, there was a netted sleeping hut on the flat roof though one always slept under an additional sleeping net up there as a safety precaution. With so much irrigated land in the vicinity, mosquitoes were very common, even though there was a constant spraying of the smaller channels and any stagnant water to try to prevent them breeding. Every now and again, the cook or the houseboy would disappear for a few days, and could be found in the servants' quarters rolled up in a sheet on his bed recovering from a bout of fever, which they seemed to get over without much difficulty, as they had presumably built up a degree of resistance from constant exposure since childhood.

Quite a lot of wives came out to be with their husbands for the winter months from September to March. Some found time hanging rather heavily on their hands with servants doing many of the household jobs that they did themselves at home, and would bring out with them for additional company unmarried female relatives or friends. These were rather crudely and uncharitably referred to as the 'fishing fleet' by the bachelor population, for obvious reasons. There is no doubt that this female influx added a great deal to the social life of the community, resulting in a marked increase in dinner parties, and other social events. The Sudan Service had the great advantage that the husbands then had their three months leave at home, often during the children's summer holidays, so that there was not the disruption to family life and separation from children, which was so often the case in the Colonial Service.

Wad Medani, in addition to being the headquarters of the Blue Nile Province, was also very close to the main base of the Sudan Plantations Syndicate at Barakat. This was the commercial farming company which managed the Gezira Scheme, covering the whole area irrigated from the Sennar dam. The Syndicate was a unique example of a successful commercial partnership between the Government, the company and the local cultivators. It worked to the benefit of all three partners, and continued to do so until the condominium handed over the control of the country to the independent Sudan Government in 1953.

Under the Scheme, the Government provided the land, which had been taken over before the First World War, with tenant rights established according to ownership at that time. It also provided the irrigation services for the whole area. The Syndicate provided the overall management of the farming operations including the primary cultivations and provision of seed at cost, and supervised the growing of the crop. It also took the cotton at harvest, managed the ginneries, and sold the cotton on the world market. The tenants were responsible for sowing the crop, for irrigating it on a 15 day rotation, for keeping it clean, and for the harvesting, and transport of produce to collecting points. All the field operations carried out by the tenants were supervised by the Syndicate officials, who were British. Other crops such as sorghum were grown, but these were the responsibility of the tenants, who kept the produce. They were also allowed to keep cattle and sheep on the land not used for cotton, though the extent of these operations was supervised by the Syndicate, to ensure a proper rotation of crops.

The profits were divided 40% to the Government, 40% to the tenants, and 20% to the Syndicate which was a public quoted company on the London Stock Exchange.

It was most efficiently managed under Arthur Gaitskill, brother of the Labour politician, and it paid good dividends, though being a one product company, it was very dependent on the world price for cotton.

This tripartite partnership worked extremely well, but, so far as I know, was never successfully copied anywhere else, except for a scheme in Tanzania run on rather similar lines in the early '60s, which was not successful.

The tenants had mostly been very small subsistence farmers when the land was brought under irrigation, and they had to be taught how to deal with new crops under intensive conditions. This was not always easy, since quite a number of them were basically cattlemen, who had to be converted into arable cultivators, and the whole philosophy of the two ways of life is different. By the time that I was in the Gezira, the Scheme had settled down very well, though, naturally, there were good and bad tenants. Some of the good ones had got quite rich and acquired multiple holdings and retired to live in Medani, employing others to do the hard manual work for them. Entrepreneurial ability will out, it seems, in all walks of life!

The coming of irrigation had brought about immensely beneficial change to the area. Instead of impoverished settlements, relying entirely on a short rainy season to provide enough water to grow sparse forage, and short-life, low-yielding crops of millet, there was guaranteed water for most of the year, and the ability to keep and feed cattle at much higher levels of nutrition. Villages were created or rebuilt, schools and dispensaries provided, and the whole standard of living rapidly improved. Not all tenants were satisfactory, since the Sudanese in this area were a proud people with Arabic traditions who tended to regard hard manual work with a certain amount of disdain. The men much prefer to sit down and pass the time of day with cups of coffee discussing local politics, while the women and children get on with the manual work. The Syndicate Inspectors, therefore, had quite a hard time in maintaining overall high standards of production. That they were able to do so depended to some extent on the system whereby each main operation carried out by the tenant was paid for only on completion of the job.

The soil of the Gezira is very uniform, but it is also very heavy, with something like a 60 per cent clay fraction, which makes it very tough to work. As with most forms of dry land farming, and in spite of irrigation, conservation of water is very important, so total inversion of the soil by ploughing is inadvisable, to say the least. So the land was merely broken up by heavy cultivators which were winched across the blocks of land by diesel engines at each side – exactly the same principle as that used with steam tackle on the heavy soils of the Eastern Counties in England in the late 19th and early 20th centuries. After breaking up, the clods were ridged and irrigated between the rows, which helped to break them down.

Cotton seed was planted in the ridge by hand, with one man of the gang going ahead with a curved stick (a seluka), which he thrust into the ridge, giving it a sharp turn as he did so. This left a small hole, into which the wife or children following dropped several seeds, and covered them over. This was done in the wet season in July, when the first rains should have softened the clods still further. After some six weeks, it was thinning time, when the number of plants per hole were reduced to two or three, the land then being ridged up again to facilitate irrigation. Four and a half inches of water was applied every 15 days until the end of March, when the crop was sufficiently advanced to need no more.

Picking was done by literally thousands of men, women and children, from the villages and towns, such as Wad Medani. The cotton, after weighing by the Syndicate staff, was pressed into large hessian sacks, and taken by camel to collecting points, and from there to the two ginneries. Once picking was finished, every bit of crop residue was collected and burned as a hygiene measure against the spread of fungal disease and the boll worm, both of which could devastate a crop if allowed to carry over from one crop to another.

This was how the crop was grown on the Syndicate areas, and we used exactly the same methods on the research farm, being meticulous in burning every little piece of leaf or stem. It was a wonderful sight to see the Gezira virtually ablaze in the evening light, and it was a spectacular end to each growing season.

Another hazard in growing crops in the Gezira, and this applied equally to sorghum, or any other leafy crop, was locusts. We had two quite small swarms when I was on the Research Farm, and fortunately neither settled properly. It is quite an awesome sight as the huge dark cloud approaches, blotting out the sun, and it was amazing how quickly the word seemed to get around the local population that a swarm was imminent. Every man, woman and child in the vicinity would come rushing out into the fields carrying every pot and pan that they could lay their hands on. As soon as the locusts were overhead, these were beaten vigorously, making a wild cacophony, with the hope that the noise created would deter the locusts from starting to land, and that they would be driven off somewhere else.

If they did land, as part of the second swarm did, there was really almost nothing that could be done, because of the huge numbers involved. They were like a green river flowing over the foliage of the crop, making a horrible crunching sound devouring it as they went, just leaving barren stems behind them. These days, no doubt, in an area of the importance of the Gezira, aerial spraying facilities would be at hand.

Socially, life was very agreeable. There was a so-called golf course just outside Medani. Naturally there was no grass on it, but one got used to playing off bare ground, and putting on oiled dead level sand greens where a putt directed straight at the hole was sure to go in. The only real hazards were wandering camels and goats and some thorn scrub. It was hardly St. Andrews, but quite good fun, all the same. Then there was tennis and squash at the social club, or riding. Personally, I didn't ride in the afternoons as I had usually been out on the horse doing morning rounds, and that was quite enough for the day. I was never much good on a horse, as I had hardly been on one before and didn't enjoy it very much. In addition to that, the pony that I had bought was a far from pleasant natured beast (perhaps that was why he was on the market when I arrived). I think that he had been trained to play polo, for every now and again, on our perambulations round the farm, he appeared to get bored, and would suddenly, for no apparent reason, dash off leaving me bouncing about on the saddle desperately trying to regain control.

On two occasions, he achieved his apparent objective of unseating me. On the first one, he took off into a mad gallop, and I managed to hang on for a time. Then in full flight, he suddenly saw an irrigation channel in front of him, and swerved violently to one side. Somewhat naturally, I carried straight on, landing upside down on the edge of the ditch with one leg in the water. Whereupon, he trotted off contentedly back to his stable, leaving me to walk back from the far end of the farm which naturally did not endear him to me. The second occasion could have been more serious, for the same sort of thing happened, but this time one foot caught in a stirrup, and I was dragged along the ground for a short distance before I got free. Apart from a lot of bruises, no serious damage was done, but this time, he did have the courtesy to stop and wait, so that I could, at least, ride back home to get cleaned up.

All this time, I had been slogging away at the Arabic, both written and colloquial, but was making rather heavy weather of it in spite of two lessons a week from a senior technician in the Plant Physiology Department, who had a good reputation for cramming people through the exam. Like many

Sudanese officials, a product of the College in Khartoum, Mahomet Tome was a charming man who spoke excellent English. He was widely read, and very knowledgeable about international matters, and we would usually spend much more than the allotted hour discussing a very wide range of topics, mostly about the country and its future. One could not help feeling that he and Awam Nimr, and several others like them, were capable of doing much more responsible jobs than were available to them at that time. It must be one of the tragedies of modern Sudan that when the time came for the condominium to end and the British withdrew, political control at the highest level fairly quickly got into the hands of Arab extremists and Communists rather than into the hands of the cultured and enlightened people that I came to know, and to respect so much.

I did eventually pass the exam, but only because of the benevolence of the examiner, I suspect. Now, 50 years on, I can hardly remember a word of the language – what a waste of time!

After his six months sick leave, Edward John returned, and I had to learn to play second fiddle again, which I didn't much like after my first taste of power. It became rather obvious that there was not really enough work for two inspectors on the farm, so I was quite glad in a way when news came from Khartoum of a new posting, which was not unexpected. It did, however, confirm my worst fears for it was to be to a place called Tokar on the Red Sea, which had the reputation of being by far the worst station in the country. Again, I had mixed feelings about moving on. On the one hand, I was ready for something new, especially as the job was not so attractive once Edward had returned. On the other, I was sorry to have to give up the house and garden and our small mess, and to leave a lot of new friends, both British and Sudanese, and the easy life of the Station.

Second Posting

So, by the middle of August, it was packing up time again. My young servant, Abdel Karim, pulled a very long face when I broke the news to him that Tokar was to be the next stop, but he agreed to come with me, as did my cook. Nearly all the servants in the Sudan came from the same area, the town of Dongola, on the Nile north of Khartoum. Exactly why this was, I never discovered. They obviously had some kind of underground bush telegraph, for as soon as a newcomer arrived or a job became vacant, there would be a brother, or a cousin or some relative waiting on the doorstep the next day applying for it. I was told that they much preferred to work in a bachelor household to those in which there was a wife around, presumably because they were generally given a much freer hand in running the house, and also, I suspect, because they were able to get away with a lot more perks. They were generally perfectly happy to move around the country with their employer, as they could be pretty sure of finding either relatives or friends working for other employers in the new location. Abdel Karim had obviously heard some pretty gruesome stories about Tokar, but the unemployment rate among servants was high, and he knew on which side his bread was buttered. I was glad that he decided to come as he was very reliable, honest and spotlessly clean in all respects – an important consideration in a country such as the Sudan.

Tokar is a small godforsaken town about 120 miles south of Port Sudan lying some miles inland from the Red Sea. Its one claim to fame is that the area produces some of the highest quality cotton in the world. This is grown on silt deposited on the delta of the river Baraka, whose waters pour down out of the Red Sea Hills in spate after the rains in July and August. They spread out over a section of the delta, leaving their silt behind as the speed of flow diminishes and in a good flood this can be to the depth of a few inches, though it will vary over different parts of the flooded area. The cotton is planted as soon as it is thought that the last flood has occurred, though occasionally there will be a late flood, and early sowings can be buried, or possibly washed away and then it all has to be resown. There is enough water in the silt (which holds it well) to germinate and grow the plants until lighter secondary rains fall over the delta round about

Christmas time. These rains should provide just enough water to mature and ripen the crop. It is thought that the quality is partly due to the high relative humidity of the area, and to its proximity to the Red Sea, though I am not sure whether this is the correct reason. In any case, the product has a long staple of excellent strength, which commands a good price on the world market.

The other characteristic of the area, apart from the sticky heat typical of the Red Sea as a whole, is the wind which blows strongly from the sea to the mountains for three to four months of the year, and then in the opposite direction for another four months or so. As only about a third of the delta is flooded each year, and it may be 20–30 years before the silt builds up high enough in one section to change the course of the flood to another section, it means that two-thirds of the delta is dry and arid, and the fine silt of which the soil mainly consists is very subject to wind erosion. This causes dust storms virtually every day during the two windy periods, which are in the autumn and then again in the early summer before the rains. Dust storms in the Sudan are known as haboobs, and in Tokar the less severe autumn blows were called the habbebais, or little haboobs. During the three summer months, by which time the cotton will have been picked, as many people as possible leave the town, as the conditions become almost unbearable with heat and dust.

During the habbebai season, the wind rises at 10.00 in the morning whipping up the dry silt into an almost blinding cloud, and this continues till 4.00 in the afternoon, when the wind suddenly drops, the dust settles, and the sun reappears for an hour or so, before it drops down behind the mountains to the west, and it is dark by about 6.30.

The fine dust seems to penetrate everywhere, even in the houses with wire mesh shutters, and closed windows, and a tablecloth which is white at the beginning of lunch is covered with a fine grey film by the time that the pudding is finished. Hence, the apocryphal story of the District Commissioner, who concerned at his consumption of dust, took to eating his lunch in his bedroom out of the chest of drawers, opening the drawer when he wanted a mouthful, and closing it again rapidly while he ate it. When outside, it was essential to wear close fitting motorcycle type goggles to protect the eyes from the flying dust particles, which meant having the damned things on for at least four hours on a normal tour of inspection in the field.

The Government, in a far-sighted move prior to the First World War, had taken over all the land in the delta, and marked it out on a grid system of defined areas, much as it had done in the Gezira, and allocated areas in each of the three sections to the existing tribes according to their resident populations at the time. This meant that if the flood did change course to another section, each tribe would still have its defined areas laid down for it. The tribal areas were occupied by individual tenants, whose tenure was recorded in the agricultural department in Tokar, so that one could always find out who the nominal tenant was, if things were not being looked after properly on a particular holding.

The main crop was cotton, grown in rotation with sorghum, beans and any other saleable crop that the tenant wished to grow. In practice, quite a few would grow some vegetables, for sale in the town, or for home consumption. Sorghum was also grown in strips round each block of cotton as a windbreak to protect the leaves of the cotton plant, which in a bad blow could be torn to shreds, in the same way that the leaves of sugar beet can be ripped to bits by a blow of soil in East Anglia. The sorghum grew faster than the cotton, and its tougher leaves withstood the abrasive effect of the dust particles better. The tenants also kept a few cattle, sheep or goats.

The movement of sand and silt was causing considerable concern as it was threatening to bury the town in one vast sand dune, and it was an unending task keeping the roads reasonably clear, and stopping the houses from being swamped by drifts. It was a dull, drab little town, with no gardens or greenery to be seen, apart from a few straggly acacia trees. This didn't matter a great deal to me, for as soon as I arrived, my Senior Inspector told me that my job was out in the cotton growing area (which was then some ten miles out of town, as the flood had come down the east side of the delta for a number of years). This meant that I spent six days a week, camping out in matting huts, which had been built at three separate locations, scattered across the area. The object was to avoid excessive travelling, in view of the distances involved. As there was only one government car, which was needed in Tokar, it meant being taken out each week and left there to do one's work either on a camel or on horseback.

The job basically was to ride round a different area from one of the huts to make sure that the tenants were getting on with whatever job was important with the cotton at the time. There were a number of Sudanese district agricultural foremen stationed in the delta, and one came round each day to take me to any holdings where it was felt that a bit of pressure from the British inspector would have a desired effect. I suppose it was necessary, but it was soul-destroying work just telling cultivators to get on with it, or they would be for the high jump.

The other part of the job, and a much more interesting one, was to survey all the watercourses after the previous summer's flood, and to build dams at strategic points so as to halt the flow of the water next year, and spread it out widely so that it could deposit its silt, instead of just rushing down and losing it in the sea. If the dams were not built, the force of the water gouged out deep channels, and a lot of valuable sediment was lost. I enjoyed this part of the work, as it was positive and constructive, and I suppose it also took one back to childhood and building sandcastles, or making dams in ditches in Epping Forest when I was very small.

The actual surveying of the places to be dammed, and also the subsequent supervision of the building was in the hands of a marvellous elderly Sudanese surveyor, Hamid, employed by the Government Surveys Department. He was quite small for a Sudanese, with grizzled hair, and a wonderful twinkle in his eye, and he had lost a leg in some scrap in which he had been involved in his youth. He hopped around very nimbly, with the aid of a stick, which he was always waving about, and a battered old wooden leg, which looked as if it would collapse under him at any moment from dry rot, or sheer old age. He spoke very good English, and we got on very well, and I was always glad of an excuse to pay a visit to the section in which he was working. He was also responsible for taking on the casual labour which we employed for doing the work, which might amount to 40–50 workers for the larger jobs. It was hard work for them in the humid heat, digging the soil out from an area downstream from the dam, and carrying it in baskets on their heads or shoulders to be dumped on the new dam. Most of them were terribly thin and under-nourished, as they could hardly ever afford to eat meat, existing largely on a carbohydrate diet. At the end of one big job, I remember Hamid suggesting that I might authorise him to buy a sheep, to give them a good meal as a reward for conscientious work, which I did very gladly, though I fear that one sheep did not do much to remedy their mal-nutrition. Four hours work a day under those conditions was about all that they were fit for, and I learned not to be too critical about what appeared to be laziness, when the root cause was almost certainly poor feeding. Sadly, I was not there to see how effective our dams were when the flood came the following year.

I did most of my work on the camel that I had acquired, alongside a new horse, within two days of reaching Tokar. It was more comfortable and much less tiring than riding a horse for the distances involved which were seldom less than 10 miles a day. I had to buy the camel, but a generous forage allowance quite quickly covered his cost. Camels are not attractive animals, and they always seem to be grunting or groaning about something, especially when it is time for them to get up and go to work. Their other unpleasant habit is to drool a nasty greenish sort of saliva out of the corner of their mouths, which looks, and probably is, extremely insanitary. Mine had a go at biting my legs on two occasions, for no apparent reason. Luckily I got them out of the way in time, as I am sure any graze would have gone septic very quickly.

Riding a walking camel is no fun, as it lurches from side to side, but at a good trot it makes a comfortable ride with a fairly gentle bumping up and down on the well-padded saddle and one does not have to grip with the legs, which makes it more restful on a longish trip. A gentle gallop is smoother still.

The pony was a much more docile animal than the previous one, and I used him rather more for short evening rides than for the morning's work.

It was a spartan existence, for the matting huts were extremely primitive, consisting of one inside room for sleeping, and an outer section, with one side open, for sitting and eating. It was warm enough all the time not to need a fourth wall. The servants had a similar hut a little distance away, for cooking and living quarters. The day's routine was always the same, with a 6.30 rise, followed by breakfast, and off on the rounds at about 8.00, returning between about 1.30 and 2.00. After a good wash to get the dust out, and lunch, it would be a siesta, followed by a cup of tea at 4.00, and then perhaps a ride to enjoy the evening sun before it got dark. Supper at 7.30, and then reading or a wrestle with the Arabic books till bedtime. The evenings were not the restful times that they should have been, since the hurricane lamp attracted every conceivable type of insect, in their thousands, all absolutely intent on committing suicide by flying into the lamp. One had to have a large bowl beneath it to catch the corpses as they fell, and this led to a rather dim religious light which had a depressing effect on the brain.

Before moving down to Tokar, I had had the good sense to buy a short wave radio set, on which I could pick up some programmes from home, though the reception wasn't all that good. But it did provide a lifeline to the outside world. In fact, my first week out in the cultivation area was the first week in September 1939, and I remember so well, as I suppose everyone must who heard it, the announcement by Chamberlain of the declaration of war. It was a strange feeling, being so completely isolated at such a moment, and knowing that one was cut off from home and quite incapable of doing anything positive towards the war effort, and it was all the worse for having no one to discuss it with. Over the following weeks, the absence of any dramatic war news was reassuring, and the sense of frustration at being away from everything disappeared as the routine of the new job became established.

There were only three British officials stationed in Tokar: the District Commissioner, the Senior Agriculture Inspector and myself. The DC was hardly ever seen, as he had a huge area to cover, and he was nearly always travelling his district, keeping things peaceful. Some of the tribes in this area are quite aggressive, particularly the Haddendoa, known for their wild appearance and their fighting qualities. There were always minor squabbles to be settled between the different tribes over grazing rights, and things of that kind. The Haddendoa are essentially nomadic in tradition, and cattle-men rather than cultivators, and though they had areas allocated to them in the delta, they were often the ones with whom we had most difficulty in getting work done. They are a very independent and proud people, and one could not help admiring them in spite of their rather villainous appear-ance, accentuated by the huge crop of fuzzy hair sticking upright all over their heads. There was always the feeling that they might suddenly bash you at any moment with the stick that they usually carried across their shoulders. Their neigh-bours did not like them much either, so there was always some friction for the DC to smooth over. Tokar is not too far from the border with Eritrea, and border incidents fell into his lap as well.

My Senior Inspector spent most of his time in the Tokar office, dealing with sales of cotton, and negotiations and problems with tenants and their payments. He was coming up towards the end of his service, and had been in Tokar for a number of years, perhaps rather too long, and I found him rather withdrawn and difficult to get through to. As I only saw him about once a week, this didn't matter a great deal, though it might have been pleasanter to have had an extrovert to talk with on the one night of the week when there was a chance to talk at all about England, the war, and things of that kind.

I had been in Tokar for about two months, when I first began to feel rather out of sorts, tired in the mornings, with a poor appetite, and generally lethargic. It was a bit like an undergraduate hangover, following a night out at which the drinks had been badly mixed.

Then, about the middle of November, came a summons to attend a six weeks' military training course (starting early in December) with the Cheshire Regiment, the Regular battalion stationed in Khartoum at the time. There were 16 on the course, drawn from different Government departments, and I found myself sharing a room with Wilfred Thesiger, one of the young District Commissioners. He had already achieved a reputation for going off on long journeys on his own and was clearly an exceptional character – a sort of cross between T. E. Lawrence and Doughty of Arabia. Ordinary comforts and conventional living obviously meant little to him, and he was already perhaps preparing for his later historic journeys in the Empty Quarter or the Marshes of Iraq, which were to bring him international acclaim, and finally a knighthood at the age of 84 in 1995. Great explorers need to have qualities of austerity and self denial if they are to endure hardship and deprivation over long periods, but it does not make them particularly easy to live with.

The major problem for me on the course was keeping up with the hard physical training and exercises as I was getting increasingly debilitated. Finally during the fifth week, I landed up in hospital. Here, there was talk of hepatitis and amoebic dysentery, but the tests proved negative, and I was posted back to Tokar, as the course was almost over by then.

Tokar in January was a rather pleasanter place than before, as the winter rains had settled the dust, and the air was clearer, and the sun shone for longer. The crops had grown at a great pace in six weeks, and continued to do so through February and March, and the work apart from dam building was trying to get tenants to control weeds. The cotton was in flower, and the crops became quite attractive to look at, so life should have been better. But, by the end of April, I had lost two and a half stone in weight, felt very weak, and found myself back in hospital, first in Port Sudan, and then in Khartoum. Tests there again proved negative, but by then, as I was in a very low state, it was decided to pack me off home on sick leave as soon as possible, particularly as the situation in Europe was beginning to look extremely threatening. The Government did not want any lame ducks on its hands when it was likely to be confronted by the Italians across the border in Ethiopia and Eritrea who might well try to take over the country to link up with their compatriots in North Africa.

So on 10 May 1940, I beat an ignominious retreat from Africa, with a most uncomfortable feeling of failure at having proved incapable of standing up to the physical and possibly mental challenges to which I had been subjected. Tokar, with its evil reputation, had again proved to be the winner.

I had been lucky to get a seat on an Imperial Airways flying boat on the regular Cape to Britain service, as Khartoum was one of the staging posts. The service had been started in the mid '30s, primarily to provide a rapid mail link with South Africa, but also to enable businessmen and Government officials to reach the main cities of Africa much more quickly than by sea or overland services. It had to rely on the availability of water, and the flying range of the Short boats which were used was restricted to about 350 miles. The full trip from Hurn, the British base, to Cape Town took a week, but that was twice as fast as the sea route, and it took four days to get home from Khartoum, compared with the eighteen days by sea and land.

There were about twenty passengers, and the flying speed was, I think, between 100–150 miles an hour, at an altitude of around 1,500 ft. Overnight stops were at Alexandria, Athens, and Marseilles, with intermediate re-fuelling points on rivers or lakes or the sea. The only tricky bit was over Egypt where there was considerable air turbulence between the desert and the cultivated areas, where we pitched and tossed about quite alarmingly. It would have been a fascinating and enjoyable experience if one had not been feeling like death at the time. It later became clear how lucky I had been to get a seat on that plane, since it must have been the last one to get through the Mediterranean before Italy entered the war – the last one, in fact, ever to make the trip, since the service was not revived after the war, by which time conventional long range aircraft had changed almost beyond recognition. It almost makes one feel part of history.

Recovery was very slow, and it took the whole summer before I put on much weight and began to feel that I could tackle a job again. I have much sympathy with those people who today are suffering from ME, for I think I may have had something very similar. The mornings are the worst with a sense of lassitude and weakness, and every task becomes an effort. One has no stamina and a vital spark appears to be missing somewhere; it is all the worse because there is no outward sign of illness. It was particularly difficult to put up with in 1940, because of the war situation following the fall of France, the entry of Italy into the war, the air raids and then the Battle of Britain and the threat of invasion. It was frustrating in the extreme not to be contributing in any way towards the war effort. I had been summoned to a medical at the Sudan Office in July, and was duly invalided out of the service on the nebulous grounds of 'debility', but as the country was completely cut off by the war in the Mediterranean, I probably could not have got back there in any case.

In September I began to make some enquiries about joining one of the services or a voluntary organisation. It became clear immediately that, with an agricultural degree, I was classified as being in a reserved occupation, and would not be allowed to join up, even if 100 per cent fit. So I then wrote to Robert Rae to see if he knew of any suitable jobs, and he referred me to James Scott Watson, who was coordinating the employment of graduates for the Ministry of Agriculture. He told me that the Executive Officer of the Surrey War Agricultural Committee was looking for an assistant, and he thought it would suit me well. So, in mid October, I went up to Guildford to meet Hugh Mattinson, the Executive Officer who had been the County Organiser of Agriculture before the war, and the chairman and Secretary of the Committee. It transpired that Hugh had been pulled out of the army in 1917, and sent out to the Sudan to help to boost the output of cotton, which was getting in very short supply at the time. This immediately established a close rapport between us, and I was offered the job on the spot. It came as a great relief to feel that I was once more able to do something useful and constructive.

CONSOLIDATION

The War Years

I started work in the first week of November and quickly realised that I had been lucky once again with the people with whom I would be working. Hugh Mattinson was a quiet, unassuming man, very kindly and thoughtful, and with a great deal of experience which he was happy to pass on. For the first week or two he took me round on his visits so that I could get to know the county and the scope of the work, all the time giving me an insight into advisory work, and the activities of the Committee, and its rapidly expanding functions. For the first few days, till I had found digs, he and his wife put me up in their house, and made me feel very much at home with their delightful family of three teenage children, so it was a very happy start in the new environment.

The Committee staff at the time was fairly small, with only one technically qualified agriculturist other than myself – Bill Cullen – who had been seconded from ICI. Soon afterwards another officer was appointed to take over all the advisory and development work on the livestock side, especially dairying. This was Winston Rowe, seconded from a post in a commercial dairy detergent firm. He was a few years older than I was, and we got on extremely well together as I gradually took over, among other things, the crop work, to dovetail in with his work on livestock. The other senior member of staff was the Secretary to the Committee – W. E. Lower – seconded from the administrative department of the county council. This was a key post, as he had a legal background, and this was of the greatest importance in determining the legal limits of the powers vested in the Committee under the Defence Regulations. The farmer members of the Committee were mostly in their late 40s, or early 50s, and were a very enthusiastic group under the chairmanship of an older farmer, Sidney Moon, who was inclined to get the bit between his teeth, and try to cut corners, in order to get things done quickly. I don't know how many times I must have heard Bill Lower say in Committee, 'Hey, wait a moment, you just can't do that, or you will land up in court. The Regulations say that you can go this far, and no further.' This would stop the members dead in their tracks, and they would slow down and come to a more sensible conclusion.

There were two other moderating influences in the Committee. One was Jocelyn Bray (later to be knighted as chairman of the Thames Conservancy), who was a land agent, and a most level-headed, experienced, and charming man. The other was Tom Sutton, the Land Commissioner of the Ministry of Agriculture, who was the Minister's direct representative, and served as the main link between the Ministry and the Committee. He could give official approval on behalf of the Minister for quite a wide range of decisions, though anything of major importance had to be referred to Whitehall. Most Committees shared a Land Commissioner between neighbouring counties, but we had one of our own. Tom was a truly notable character, a qualified land agent who had spent most of his career in the Lands Branch of the Ministry, including service there during the First World War, and he was steeped in experience of the workings of Whitehall. He had retired some five years earlier to Chobham and was recalled to look after Surrey. He was about 70 years of age, and was the absolute epitome of an Edwardian gentleman. Very tall and dignified, he wore old-fashioned clothes, with a high stiff white collar and a cravat held in place with a gold pin. He was quite the most courteous man that I have ever met. He was precise and everything he said was to the point.

I worked very closely with him for nearly five years on land inspections. As he did not drive, I acted as his chauffeur, and in that time became very attached to him, and to his wife, who was equally old-fashioned and kind, and who sustained me with countless cups of tea over the years, after I had delivered him home safely from our outings.

I owe a very great deal to Tom for what he taught me about the workings of government and administration, and particularly about the importance of good manners and courtesy in public life. When the time came to move on at the end of the war, it was a very sad parting, and I like to think that both Tom and his wife were as fond of me as I was of them.

From the start, Hugh Mattinson had insisted that I should attend all the weekly meetings of the full Committee. This was so that I could stand in for him if necessary, and also get a full insight into its works, and the rapidly growing scale of its operations, and also present reports on matters in which I was involved. So this is perhaps a suitable point at which to look back at the general functions of these Committees, which played such a vital role in keeping the nation fed over the six years of war.

The Workings of the War Agricultural Committees

On the declaration of war in 1939, the emergency plans drawn up by the Advisory Committee which had been set up in 1937, were put into effect immediately. The main objective was to set up a Committee in each county which would act as the agent of the Minister in seeing that Government policies were put into effect as efficiently as possible. The constitution of the Committees and their chief officers had been provisionally settled the previous year, so that there was little delay in bringing them into action. The membership consisted principally of prominent farmers drawn from different parts of the counties, and representing, if possible, different types of farming. In addition there were representatives of the landowners, Farm Workers Union, education, and of organisations of local importance if appropriate.

The main Committee was kept quite small so that it could act in a truly executive capacity, and not get bogged down in detail. Then under the main Committee, there were a number of district committees whose job it was to put the policies into effect at ground level. These were chaired by prominent local farmers. Each District Committee was serviced by a District Officer, who took his orders from the Chief Executive Officer of the main Committee, and he was a key person in many ways. He had to organise his local committee members to get out onto every farm in the district to get land scheduled for ploughing up, and he had to provide a great deal of information to head office about the farms in the area, so that rationing schemes for fertilisers and for feeding stuffs could be brought into effect. He was provided with an office and clerical staff, but had to work very hard indeed in the early days in getting to know his district, and also be ready to transmit information directly down to his farmers, and answer their queries as effectively as possible. District Officers were generally drawn from the staffs of the colleges – where there was one – or from commercial firms, or they might have previously been NFU officials, or recently retired farmers who wanted to play an active role in the war effort. The Chief Executive Officers of the main committees were also drawn from varied backgrounds. They were chiefly the former county organisers, where the county in question had employed one, but some of them were drawn from the staffs of the universities – both of my former Cambridge teachers, for example, Jem Sanders and Frank Garner, became Executive Officers for a few years in the early part of the war.

As the war went on, and the work increased enormously, specialist departments had to be set up, with officers employed to run them, and many of them required assistants as well because of the work load. The feeding stuffs rationing scheme needed a separate unit, and the Machinery department required a Machinery Officer to oversee contracts, procure machinery, employ operatives and run the depots. Similarly, a Labour Officer was needed to run the contract labour gang service, and especially to provide labour for the farming of the land in hand, and, in due course, to organise the prisoner of war labour gangs. Horticulture required its own Committee in a county like Surrey, and had its own Chief Horticultural Officer with assistant district officers. As technical advice and demonstration work assumed greater importance to aid efficiency in production, separate Technical Development Sub-Committees were formed in most counties, and in due course, this became one of my responsibilities. All these Sub-committees reported back to the central Executive.

Naturally all this didn't happen at once, but grew during the first three years of the war, till by 1943, the committees had become very large administrative organisations indeed. I suppose my own appointment was symptomatic of the pressures, as by the autumn of 1940, when I was taken on, Hugh Mattinson was finding that the workload was more than he could manage, and he needed someone on whom to offload both the demands of the technical advisory work, and also the taking over and management of the land being farmed directly by the Committee, which quite rapidly became my two chief preoccupations, alongside a number of other smaller commitments.

The powers of the Committees were sweeping, in that farmers could be ordered to carry out a large range of farming operations. The main requirement was the ploughing up of old grassland for arable which also included carrying out the full range of cultivations and sowing to specific crops. Committee members and officials had powers to enter farms to ensure that orders had been complied with. When orders were not carried out, and the farming standards were very low, farmers could be put under supervision and subjected to frequent visits by district committee members or officials. If there was still no progress it could mean an official Executive Committee visitation, and a possible recommendation to the Minister of Agriculture that the tenant or owner-occupier should be compulsorily dispossessed of the land. In the case of a tenant the owner had to be brought into the picture, but owners were generally quite satisfied with dispossession, since it was a good way of getting a tenant out who was letting a farm get into disrepair. Dispossessions were not common, but in some cases where farmers were quite incapable, for one reason or another, of raising production, there was little alternative. A system of appeals against such drastic action was laid down, including interviews before the Executive with legal representation, and final appeal to the Minister himself, and to the local MP, with the threat of a question in the House. Appeals were rare in my experience, and in most cases, dispossession was to the ultimate benefit of the farmer himself, as it could save him from a steady drift towards bankruptcy, which would have been virtually inevitable otherwise.

One of the reasons why official orders were used in all cases for the ploughing up of old grassland was to protect tenant farmers from possible claims from their landlords under old tenancy agreements for breaking up pasture. This would have been unlikely under the circumstances, but orders under the Defence Regulations over-ruled such rules in tenancy agreements, and this was why, later on, we began to issue orders for reseeding.

There is little doubt that the reason why the system worked as well as it did, and that there was so little friction, was that the district committee members who did most of the scheduling were themselves farmers, well known and respected in their areas. They spoke to farmers in their own language and, in many cases, were able to give helpful advice about reorganising the system to cope with less grassland, or advise on arable cropping. It was rare for those who received such orders for ploughing or cultivation to claim that they were being persecuted by petty officials waving big sticks, though naturally, as in any group of people in any walk of life, there were one or two difficult characters – professional dissenters – who made such claims. They were generally not very good farmers themselves, and in the end they had to toe the line or run the risk of prosecution and substantial fines, quite apart from earning the opprobrium of their fellow farmers. The vast majority of the farming community was intensely patriotic, and only too anxious to do everything they could to try to increase food production – another reason why the system, in general, worked extremely well.

The first task of the Committees was to try to reach the quotas set by the Ministry for ploughing up grassland for the harvest of 1940, which in total amounted to two million acres. As soon as the district committees had got their farms listed out, the members were out trying to encourage farmers to volunteer at least one grass field for ploughing up and cropping – preferably with cereals. Of course, the declaration of war had come rather too late to get much new land cropped with autumn corn, so much of it had to go to spring barley, or potatoes, or possibly spring oats.

The Ministry had clearly set out its food policy, which was to get as much land as possible cropped with wheat and potatoes, the energy foods, and the other priority was to be given to milk, as a protective food. These were to be given first call on fertilisers and feeding stuffs when rationing came in later on. Second line priorities were sugar beet, and oats and barley for livestock feeding. Sheep, beef cattle and pigs got no preferential treatment and pig numbers dropped very quickly.

In spite of having at least two years warning of possible hostilities, the Government entered the war with low stock-piles of food and fertilisers, so it was important to get as much land under crops as possible, in case merchant shipping came under serious attack. This fortunately did not happen during the first year, as the Germans had not by then built up a very large U-Boat fleet, perhaps because Hitler had not believed that Britain would declare war quite as soon as she did. The real crisis in shipping was to come in 1942 and 1943, by which time, some five million acres of land had been added to the cropping acreage.

The potentiality for increasing home food production was certainly there as far as the land was concerned. By 1939, an area roughly equivalent to the size of Wales had disappeared from agriculture compared with 1870. Only some nine million acres were under crops, of which five million were in cereals, and there were 18 million acres of old pasture and a further 16.5 million acres classified as rough grazing, mainly in the hill areas of the West and North. These were depressing figures, but even worse was the fact that nine million tons of animal feeding stuffs were imported annually, and 75% of the wheat used for bread was also imported, while 45% of the barley was of foreign origin. In all, Britain was only producing at home some 35% of her total food consumption. This was a dreadful indictment of previous Government policies for a country with such natural advantages for a diversified and productive agricultural industry.

So there was plenty of old pasture waiting to be ploughed

up, but that was not the end of the story by any means. Two other major problems had to be faced. The first was a very grave shortage of equipment for doing the job, and the second was the complete lack of arable experience of some farmers – often those with the most grass.

At the outbreak of the war, there were only about 50,000 tractors in the country which meant that the majority of farmers did not have one at all. There was still, it is true, a large population of working horses, since the numbers of these had not dropped commensurately with the increase in the number of tractors, but many of these were not trained for ploughing and arable work. Furthermore, the number of ploughs, cultivators, corn drills and binders were inadequate to cope with the demands made by an extra two million acres of cropped land. The difficulty was not so much getting the acreage for cropping but in knowing how best to advise those farmers who had no equipment or expertise of how to deal with it.

Ultimately the problems were met in three ways. Firstly, the larger and more experienced farmers who had already got some equipment came in and did a lot of work for their neighbours on a contractual basis. Secondly, the Committee had to set up its own machinery depots and equip them so that they could take on a proportion of the work on contract. Thirdly, after the first year or two, many of the larger farmers had been able to make sufficient money to buy some equipment for themselves. Tractors still remained a problem for some years, for the demand for them always exceeded the supply, in spite of a tremendous effort by the manufacturers – Ford, in particular – to increase production. Considerable numbers were imported from the United States, though quite a lot unfortunately went to the bottom of the sea in the dark days of 1942–3. The demands on the Committee's resources in Surrey were so great, that the initial two machinery depots had to be quickly increased to six.

Another of the functions of the Committee was to allocate its county quota of new tractors as fairly as it could between farmers (who had to make specific cases for them) and its own depots. This was not an easy job, as there were not enough to go round. Inevitably, some of those farmers who did not get one immediately were very critical of the Committee, but farming some 7,000 acres in hand on top of the contract work placed a tremendous load on the shoulders of the Committee depots. Nationally, the figure of 50,000 UK tractors increased to 95,000 by 1942 and to 140,000 by 1945, with Ford producing no fewer than 18,000 in 1941, which was a tremendous effort in view of the competing claims of the armaments industry.

To help with the labour situation, the Women's Land Army had been set up in 1939 on the lines of the similar body established in the First World War in 1917. It was to play an increasingly important role as the war went on and all non-essential labour was called up for the services. Mostly, girls were placed as individual workers on farms, but in areas where there was a serious need for contract labour – and Surrey was one of them – hostels were established generally quite close to the machinery depots, and the girls worked in gangs under the control of the Committee's Labour Officer. They were made available for doing contract work for farmers, for horticulturists, and for the jobs on the land farmed by the Committee, especially for the potato and cereal harvests. In addition, the Labour Department had a number of male gangs for things like land drainage and ditching.

By the time that I arrived in the autumn of 1940, most of the basic Committee structures were in place. Demands for its services were increasing rapidly as a result of the second round of scheduling for ploughing old grassland, and farmers began to appreciate that they would need more assistance. At the same time, the Committee was having to take over a considerable acreage to farm in hand, and much of this required very extensive clearing and drainage, before it could be brought into cultivation. Rationing of fertilisers and feeding stuffs was about to be introduced, and new administrative units had to be established to deal with these schemes, and also with land drainage.

After my first few weeks I began to take on specific jobs

almost by default. The major one was the requisitioning of land and the supervision of the farming of the land already taken over.

One of the chief complaints levelled at the district committees was that they were asking farmers to plough up grassland but were doing nothing about building land, commons and particularly golf courses, which were all numerous across the county as a whole.

There were still many areas of uncultivated land which had been bought for building, but had not been developed, and local farmers felt aggrieved that they should be asked to make further sacrifices while such areas escaped scot-free. The Executive Committee realised that it was open to criticism on this score, and that it would have to face up to taking over many of these areas and probably farm them directly. This was because local farmers had quite enough to cope with on their own farms.

So one of my first major jobs was to go round to inspect all such sites reported to us by the District Committees, whether it was building land or open common land. Under the Defence Regulations, the Committee had the power to requisition land with the consent of the owners, on the understanding that it would be handed back to them on the termination of hostilities.

Sometimes, it was very difficult to find the actual owners, and in a number of cases they had already gone bankrupt. Under those circumstances, notices were posted on the land to say that it had been requisitioned, and occasionally someone would emerge from the woodwork to say that the land belonged to them.

When the land was in reasonably sized blocks of 30 acres or more, as a number of them were, on the chalk soils of the south London fringe, there were no particular problems involved in farming them – apart from damage to crops from the inhabitants of nearby housing estates, which really was not too serious on the whole. But when the parcels of land were about four or five acres, and quite divorced from local farming they were an awful nuisance, unless there were other sites in the vicinity. We tended to ignore anything smaller than five acres as not being worth the trouble and expense of dealing with from a depot maybe 20 miles away.

It was my responsibility to assess the potential of the land in every case, and write a report for the Committee with a recommendation to requisition if I considered it a feasible proposition – which in most cases I would do. The next step was to take Tom Sutton out, who would forward the report to the Ministry with the recommendation.

By the end of 1943, we had taken over more than 100 sites of this kind varying in size from about five to 100 acres.

Having dealt with building sites I was instructed by the Committee to survey all the golf courses in the county to see if there was any land suitable for ploughing up, or which could be used for grazing by sheep. I pointed out that they should not be too optimistic about finding a lot of ploughable land here, since good golfing country is usually singularly bad farming country – which is why a lot of the best courses are situated where they are. That might not be quite so true today after the golf course construction boom of the last few years, when much useful farming land has fallen prey to vast earth-moving equipment, and hills and lakes have appeared on new courses from nowhere.

There were some 40 courses to deal with and I was met with a good deal of trepidation by some of the club secretaries who feared that they might be vulnerable if the course was on former agricultural land. It helped to ease the tension and establish some rapport when they found that I could speak their language and had played off a low hand-icap before the war. A few of those on the better soils had already made arrangements with a local farmer to graze sheep, and we regarded that as a suitable use for the land. On others, it proved difficult to find large enough areas of land free of trees and bunkers and undulations to make it really worth while doing anything. One certainly could not justify taking so much land as to put a club out of business, since the compensation would be enormous, and in any case, it could claim with justification that it was providing recrea-tional facilities for a hard-working public, and military personnel.

At the end of the exercise, I had found just over 300 acres for ploughing up spread over about a dozen courses, with the largest area amounting to some 40 acres, which eliminated six holes from the eighteen. I felt rather sorry for this club, when so many others had got off, but it had the misfortune to be on reasonable quality land, and to be free of trees and other hazards. In retrospect, I don't suppose that the amount of food that we finally grew on these areas made a very significant contribution to feeding the nation, but from the aspect of public relations, and making farmers more willing to offer additional land for ploughing on their farms, it was undoubtedly a beneficial exercise.

When carrying out the golf club inspections, I was also keeping an eye open for other areas of uncultivated land, particularly commons and heathland. Many of these proved to be quite unsuitable, which was probably not surprising since the reason that they *were* common land was because they were considered worthless at the time of the enclosures, and were left as common grazings. Two areas, however, which had previously been mentioned in the Committee, did appear to offer considerable possibilities, and both of them were adjacent to golf courses that I was inspecting. These were respectively Richmond Park and Walton Heath.

It was difficult to get into Richmond Park, because of military installations, but I finally got a permit, and had a good scout round with the soil auger. Most of it was clearly useless, but one good flat green area was the polo ground, which had not been used for some years, and I managed to find two other sizeable areas, which looked feasible. They were obviously very light in texture and acid, but I deemed them capable of growing potatoes, oats and barley. It seemed sacrilege to suggest ploughing up a polo ground, and I thought of all the horsey aristocrats turning in their graves – and possibly living ones who might make a fuss. But all is fair in love and war, it is said, so I recommended the requisitioning of 750 acres in three blocks. But there was one big problem – the deer, which had been there since Henry VIII or earlier still, and could not possibly be disposed of. Negotiations with the Parks Department were protracted, but agreement was reached providing the Committee erected a deer-proof fence round the ploughed areas. It must have cost a large sum of money, but we did get four years cropping from 750 acres, with quite reasonable yields, especially from the polo ground.

One would never get away with something like that in this environmental age, since it would be claimed that the character of the plant environment would be destroyed for ever. Not a bit of it, for driving through the park recently, it was quite impossible to see where we had ploughed. Time heals rapidly when nature is left to itself.

The other area, a more dubious proposition, was a large stretch of common land, covered with scrub, adjoining Walton Heath. I found quite a fair depth of soil, and, on the basis that if it grew good scrub, it might also grow reasonable crops, recommended taking over some 550 acres to clear and crop. It was quite exciting and rewarding seeing a whole new stretch of virgin land in place of a mostly impenetrable mass of thorn and gorse scrub. In case I am accused of environmental vandalism, we did leave a very large area of common land undisturbed on the slopes. The soil turned out to be very variable and patchy, some very acid and phosphate deficient, but yields were not too bad by mid '40s standards. Disaster struck in 1944, when about 100 acres in potatoes went down with a virulent attack of potato blight, which was not spotted till it had gone too far to spray with Bordeaux, the only treatment in those days. By the time I was alerted, the beautiful crop that I had seen a couple of weeks earlier was just a mass of black rotting haulm. It well illustrated the hazards of Committee farming where large numbers of isolated areas of land were farmed by remote control. It was quite impossible to keep a close eye on it all, and the machinery officers and tractor drivers were naturally not trained to look out for technical problems.

For the first two years of the war, most of the land taken over by the Committee consisted of the types of vacant land that I have described, but as hostilities continued, and the food situation began to deteriorate, the screws began to be put more strongly on farmers who, for one reason or another, were not pulling their weight. All farmers had already been classified into three grades – A, B and C – by the District Committees on the basis of their competence and output. Those in the C category were put under the supervision of the District Committees. Often a specific member of the Committee was given the responsibility of trying to help and advise the farmer concerned.

If there was no subsequent improvement and the occupier did not respond to orders from the Executive, the Committee could recommend dispossession of either the whole of the holding or part of it. If this was done, the Committee could then either let the land to another tenant on a short term basis – which for obvious reasons was not a very attractive proposition – or take it in hand to farm it direct. In such cases, and thankfully there were not many of them, it fell to me to visit the farmer and the farm, and to write the report to be sent by the Executive to the Minister, recommending dispossession. I disliked this job intensely, for in nearly every case, one could not but feel very sorry for the individual concerned. I felt that it must be a frightful indignity in the case of an elderly man, who had perhaps pottered along for years milking a few cows and accepting a very low standard of living, to be branded as incapable of managing his farm properly, and possibly to lose his house as well. We had one or two distressing cases, too, of younger men who had taken over a farm on the death of a dominant father who had never allowed them to take any management responsibility whatsoever. They had probably left school at 14 to go back to work on the farm as virtually unpaid labour, and been given no opportunity for training. Suddenly left on their own, they had neither the technical knowledge nor the experience to cope with the everyday management of the farm and were left floundering, completely out of their depth. Something that has contributed greatly to the post-war development of the agricultural industry has been the willingness of farmers to send their sons away to get outside experience and technical training.

The reports were not at all easy or pleasant to write, as one had to stress the deficiencies in management, and could not really go into extenuating circumstances or the hardship angle. I always consoled myself that in practically every case I had to deal with, the individual concerned would be better off, and ultimately probably happier, to be relieved of the stress and worry to which they were subjected. In the case of an old man, he might be able to get out with just enough money from selling up to retire to a small cottage and perhaps get a part-time job, and in the case of a younger man, there were heaps of opportunities for employment in wartime. If left to go on, they would almost certainly become insolvent, which would leave them with nothing.

I nearly got the Committee into serious trouble with one of my reports by erroneously stating that a shortage of capital was one of the reasons for the low standard of the farming. The individual concerned was a young man who had been put into the farm by his father, with little experience, and the farm had got into a real mess. The father wrote to his MP, who was sent a copy of the report by the Ministry of Agriculture, and the father hotly denied that there was any shortage of capital involved. We were threatened with an imminent question being asked about the case in the House of Commons, which necessitated a rapid climb-down. I learned then how nervous civil servants can become when faced by the threat of a question in the House, especially when they have not got a good answer to give to their Minister in reply! It also taught me a lesson as to how important it is to be absolutely certain of one's facts before committing them to paper, especially when civil rights are involved.

I got into another spot of bother when requisitioning a large area of building land on the south London fringe. I always took out the 25 inch Ordnance Survey map when inspecting these sites, but some of them were many years out of date. It was late on a very cold and foggy afternoon when I finally located the site, which was on a slope. I walked the lower boundaries, and across the middle of the field, but as time was going on, and it was going to be dark early, I didn't go right to the top boundary of the land, which was in the fog. When the owners – a large building company – received the requisition notice, accompanied by the plan of the field, they were naturally rather surprised to find the Committee were proposing to requisition a double row of houses which had been built along the upper boundary of the field just before the outbreak of war. They expressed their dislike of the proposal in strong terms and, once again, a climb-down was required and a suitable apology made, which fortunately was accepted with good grace.

I had a much more alarming experience, which could have had extremely unpleasant consequences, during an inspection of a large, completely derelict field which is now probably incorporated into Gatwick airport.

When I was close to the centre of the field, there was a furious barking of dogs, and before I knew what was happening, I was surrounded by five snarling, emaciated greyhounds with bared fangs, snapping at my legs. My only protection was two closely rolled 25 inch maps. I am certain that if the dogs had drawn blood they would have done their best to get me down and taken a good meal. Deciding that attack was probably the best form of defence, I set about beating them over the head with the rolled maps, shouting horrible obscenities. The maps kept them away from my legs while I retreated backwards as fast as I could, taking great care not to trip up in the long tussocky grass. Mercifully, they finally called off the attack.

COMMITTEE FARMING

The acquisition of land continued apace at the rate of some 2,000 acres a year from 1941 to 1943, and the scale of the farming operations increased commensurately. The actual work on the land was carried out by the ever burgeoning Machinery Department, working from its then six depots. Hand operations were mainly done by the Land Girl gangs, or, after 1943, by Italian prisoners of war, and then by German prisoners who were generally much more hard working. Without any specific appointment from the Executive, I gradually assumed full control of the technical side of the operations, co-operating with the heads of the Machinery and Labour Departments, with whom I fortunately got on very well. We planned the cropping and management of the land, and I kept an eye on the progress of the crops as far as possible, and investigated any crop failures. In the last resort, if I was stumped technically, I called on the services of our regional advisory soil chemist at Wye – Norman Pizer. He was a first class adviser, and proved to be a great help when we had problems. We used quite a number of the sites for putting down demonstration plots for the benefit of local farmers, and this fitted in well with my other main job, which was the control of all technical and crop advisory work, under the Technical Development Sub-Committee which was set up in 1943.

To demonstrate the scale and complexity of the direct farming operations, I cannot do better than to quote from the annual report that I wrote for presentation to the Cultivations Committee in June 1944:

> The farming activities of the Committee began early in 1940, with a few acres of building land and two tractors based on Guildford. The area increased so rapidly that it was found necessary to decentralise in 1941, and the county was divided into four areas with a main depot in each. These were then increased to six areas in 1943. By the summer of 1943, land taken from commons and parks amounted to 1,200 acres, and from golf courses to 320 acres, and for the harvest of 1943, a total of 2,600 acres were in wheat or rye, 600 in oats or barley, and 675 in potatoes. Yields on the whole were good, especially in the case of wheat.
>
> For the 1944 harvest, approximately 3,200 acres are in wheat, 550 in barley, 700 in oats and 750 in potatoes, and there is a large area in seeds leys and fallow – 7,000 acres in all. The complexity of the farming operations can be gauged from the fact that this area has to be farmed on no less than 170 sites varying in extent from 2 to 350 acres. The maintenance of fertility is difficult, especially as there are few buildings, or resident workers, so that keeping livestock presents problems. On the light land, sheep have been folded for the past three winters, with a maximum of 2,300 lambs

in 1942–3. Four large straw yards have been put up and where available some buildings modified, with 230 head of horned stock wintered in 1943–4. At Limpsfield, buildings have been modified for milking, and 20 dairy cows are kept for training Land Girls in milking.

Contract Work

A large amount of contract work is now done for farmers, which in 1943 amounted to 22,000 acres, with 6,000 acres for ploughing, many for small areas. The grassland reseeding scheme has increased the contract work, with 560 acres reseeded for farmers this spring. To deal with this work, the Committee has now nearly 200 tractors, which are kept fully employed.

This brief review has been presented, as there is sometimes criticism if implements or services are not immediately available for hire at the depots, and this criticism might be due to an insufficient appreciation of the magnitude and special difficulties of the Committee's farming operations.

No additional land was taken over after 1944, as most of the acreage that it was feasible to bring into cultivation had been absorbed by that time and, with the better news from the war zones, it seemed as if the tide was turning, and the food position might not be quite so acute in the future. In fact from 1945 demands began to come in from some of the building companies for land to be released again – and this brought in a completely new aspect, namely, issues relating to the future use of land, and whether it should be developed or not.

What the whole cost of these farming operations must have been, one shudders to think, but bearing in mind the critical food shortages of 1942–4 at the height of the U-boat campaign, the produce from 3–4,000 acres of wheat, and 6–800 acres of potatoes must have made a significant contribution each year. There is no doubt at all, either, that the Committee's farming of such a large area of non-farming land did have a very salutary public relations effect on the work of the District Committees, and helped them to get their quotas of land for ploughing each year from the farmers in their areas.

As the need arose, sub-committees were set up to cover routine aspects of the different services before they reached the full Executive, each of them having one or two main committee members sitting on them. The Cultivations Committee had been set up very early on in the war, with the responsibility for vetting and issuing all the ploughing and cultivation orders which came in from the districts, and then as the acreage of directly farmed land grew, this committee also assumed responsibility for the cropping of that land. So I became more closely involved in its work, and took over the vetting of orders from Bill Cullen. This was to ensure that each one was properly worded, and did not contain anything that it might be impossible to carry out, e.g. a sowing date that could not be achieved. This was important as there were always critics ready to seize on silly mistakes, which might make the organisation appear ignorant or inefficient. In the early stages this committee also dealt with technical matters and advisory work. But by 1942, this had begun to assume much greater importance, and a new Technical Development Committee was set up, and it became my responsibility to service this one as well.

A Return to Grass

By early in 1943, Hugh Mattinson was getting concerned about the longer term maintenance of the fertility of the land which had been ploughed out of grass in the first year or two of the war, and which could already be in its third cereal crop. I well remember him asking me what I thought about it as we returned one day from a meeting. He felt that we ought to be starting to introduce a reseeding policy. It has to be remembered that the fertiliser rationing scheme only allowed for 1.5 cwt of superphosphate per acre for cereal crops, and that much of the old pasture ploughed up was fairly low in phosphate in the first place, as it had been siphoned away in milk or bones during the years of depression. Potash was only allocated for root crops, and for a few special crops or deficient areas. Nitrogen was not rationed, but was not used in large amounts, so there was a considerable drain on fertility on farms where cereal production dominated the system.

After discussion in the Cultivations Committee and the Executive, it was agreed that District Committee members should sound out farmers about their willingness to reseed if they could offer more grass for ploughing in compensation. It was agreed to stop short of issuing specific orders till we could gauge the reaction in the county. Anyhow, we were not too anxious to stimulate a sudden big demand for reseeding, because we anticipated that it might lead to a surge in requests for contracts to do the job, and the machinery depots had not got much equipment at that time for reseeding.

The response from farmers was very encouraging, as many of them welcomed the opportunity for getting some new and hopefully more productive grass to help to eke out the meagre supply of concentrates under the rationing scheme. But it wasn't always possible to find an equivalent acreage to plough up in its place. In the event, 1943 proved to be the peak year for the arable acreage in the county, since from then on, rather more land went back to grass than was ploughed out. The arable acreage that year equalled that of 1870, though, of course, relative to the total acreage of farm land, it constituted a higher proportion than in 1870 because the building boom in 20th century Surrey had taken huge areas of land out of agriculture.

Hugh Mattinson's foresight in getting the reseeding ball rolling when he did was justified, as signs of overcropping were certainly beginning to appear by 1943, and would undoubtedly have got progressively worse, if some action had not been taken then. The next question was whether we should begin to issue compulsory orders for reseeding on farms in the following spring. But before we had reached a decision, a disastrous event occurred. I had noticed for some time that Hugh had not been looking fit, and that he appeared to have lost weight, and one Friday in early November he did not come into the office and on the Saturday, I heard that he had got flu. I thought no more about it, but on Monday morning came the dreadful news that he had died on the Sunday evening, when his heart apparently just gave out.

It was a devastating blow to all of us, coming completely out of the blue, and especially so to his widow and three teenage children, a boy and two girls, then aged 13, 15 and 16. He himself was only in his mid-fifties – a very dangerous age, it seems, for heart problems.

I felt Hugh's death very acutely, for I had formed a great affection for him over the three years that we had been together, for his quiet and unassuming personality made him a delightful colleague to work with. He was a thoroughly sound agriculturist and a first class adviser who was universally popular in the agricultural community. There was tremendous respect for his judgement in the county, and this was very apparent by the way in which I was welcomed in my early days, as soon as I said that I was his assistant.

In return, I hope I provided him with someone he could trust to take some of the workload off his shoulders, and whom he found it pleasant to work with. I think this must have been the case by the responsibility that he gave me right from the start of our relationship. I can do no better to illustrate my feelings at the time than to reproduce the short obituary tribute that I wrote for *The Times*, which they published shortly after his death.

MR J. H. MATTINSON

Mr J. H. Mattinson, chief executive officer of the Surrey War Agricultural Committee, died at Guildford on November 5.

He was educated at Wye College, and was one of a selected few who were taken out of the Army in 1917 and sent out to the Sudan to organize corn production in the Northern Province. Many of the sites selected by him for Nile pump schemes have now been turned over once more from cotton to wheat growing. Returning to this country shortly after the last war, he was appointed the first agricultural organizer for Surrey, and held the post until 1939, when he was seconded as chief executive officer to the W.A.E.C.

A correspondent writes:-

The death of J. H. Mattinson adds still another name to the list of those who have unobtrusively given their lives far behind the firing line. Four years of continuous overwork and worry had so undermined his strength that he was unable to resist an infection that should not have proved fatal. Over 20 years of advisory work in Surrey had earned him the universal respect and affection of the farming community, and this affection has in turn been felt by all who have worked with him in the last few years. His knowledge of the agriculture of the county was encyclopaedic, and his brilliant capacity for forecasting the trend of agricultural policy was largely responsible for the success of the food production effort. By nature he was a man of few words, but once the reserve had been penetrated one could find beneath a deep fund of wisdom, ever at the disposal of those who sought for it, and a quiet simplicity of character which made him universally beloved.

His death left everything in limbo so far as the technical and field staff were concerned, and it was obvious that it was going to be quite a few months before a successor could be in post. Bill Lower, the Secretary, was able to take on a few of the executive jobs and decisions, but was quite unqualified to play any part in the direction of the technical staff and the District Officers. Bill Cullen, though senior to me, had been increasingly taking a back seat for the past year and did not wish to take on any further responsibility, so although I was still only technically an assistant to the Executive Officer, it seemed that I was best placed to take on the role of heading up the technical operations. So without any definite directive from the Executive Committee, but with the tacit agreement of the Chairman, I more or less assumed control. Nobody seemed to mind, and appeared happy to accept the fact, particularly the District Officers who seemed glad to have someone technically qualified at headquarters to whom they could refer if they had any problems.

The most pressing decision to be taken was that relating to reseeding, for it was already well into November and, if any official directions were to be issued, it would have to be done quickly, to give farmers enough time to prepare for it before the spring – which was the usual time for reseeding locally. The Chairman agreed that I should get the District Officers in for a meeting to get their views on compulsory orders, and also to put them in the picture as to the policy of the Executive. I also wanted to establish, if possible, a feeling of team spirit with them in case it was going to be a long time before we got a new Chief Executive. It was important to make it clear to them that it wasn't a rudderless ship, solely in charge of administrative staff.

There was strong support from the officers for a definitive reseeding policy, and for compulsory orders to be issued, so as to protect farmers from any possible come-back from their landlords. There was also strong support for full cropping programmes to be provided by farmers, with the object of helping them to plan ahead, both for cash crop production and also for the feeding of their livestock as much as possible from the farm. (Shades of IACS!)

Maps of each farm were available from the so-called Doomsday Book Survey which had been carried out in 1941, in which the District Officers had visited each farm and outlined the boundaries and the use to which each field was put in that year.

So a map of each farm could be copied, and sent back to the farmer with the cropping form to be filled up, both with the current year's crop plan, and the projected plan for the next year, including reseeding, and fresh grassland for ploughing up. Winston Rowe prepared an outline of the number of acres of feed cereals, roots, protein crops and grass required per cow equivalent so that everything was made as easy as possible for the farmer to fill in. If they were in difficulties over it they were told to contact their District Officer for help. Winston and I prepared a letter to all farmers explaining what it was all about, which went out over the signature of the Chairman of the Executive.

SURREY COUNTY WAR AGRICULTURAL EXECUTIVE COMMITTEE

" Elgin,"
London Road,
Guildford.

December, 1943.

Dear Sir,

Cropping Programme

The progress of the war has necessitated the ploughing up of an ever increasing acreage of old grassland during the last four years, and it is anticipated that still more land will be brought under the plough in 1945 and 1946. If this is to be obtained without disorganising the production of the farm and the lowering of its fertility, it is essential that a fair proportion of the old arable land should go back to new and productive grass after it has been through a course of arable cropping, and this new grass will enable a greater head of stock to be maintained on the farm, and thereby increase the amount of manure produced. It is becoming more and more important, therefore, both from the cropping and the stock point of view, that planning of the farm should be carried out well in advance, and the Committee have resolved that a detailed cropping programme is necessary for each farm.

In order to complete this task more quickly, and to enable you to plan your own cropping, and to avoid unnecessary visits of schedulers and staff, the Committee enclose herewith cropping and stock proposal forms, which you are asked to fill up and return WITHIN 14 DAYS, using the envelope in which the forms are received and the economy label provided herewith for the purpose. It is *absolutely essential* that these forms should be returned expeditiously.

The Committee consider that provision should be made first for the requirements of the dairy herd and other stock, and the following simple table should enable you to work out these acreages without difficulty :—

Stock Requirements Table

Crop	Per 5 cows or 8 young stock or 7 other cattle	Crop	Per 5 cows or 8 young stock or 7 other cattle
Hay (a) Meadow ... ... or	4–5 acres approx.	Oats	2–3 acres approx. Add 2 acres per horse or 1 acre per 15 arable sheep
(b) { 1 year ley / Pea and Oat Hay }	2½–3½ „ „		
Roots Kale	½ acre approx.	Peas or Beans	1 acre approx.
Mangolds	¼ „ „	or	
Catch Crops	¼ „ „	Arable Silage ... ... 1 „ „	

If Peas and Beans are not grown, increase Oats or arable hay to higher figure.

These figures are not to be regarded as standard for every farm, but are intended as a rough guide for an average farm, and should enable you to calculate your requirements, having regard to the conditions on your own holding.

The Committee consider, secondly, that it is essential for the maintenance of fertility, that at least 20 per cent. of the farm area should be either in one or three year leys or in directly reseeded grassland, and you are asked to plan accordingly.

A map of your farm is enclosed, and the Committee desire that this should be returned with the completed forms.

If you have any difficulty in completing the forms (for which detailed instructions are given overleaf) you should get in touch with your District Officer, who will be very glad to give you any assistance you may require.

Yours faithfully,

SIDNEY H. MOON,
Chairman.

Looking back at this, 50 years later, I am astonished about a number of things. Firstly, the rather peremptory tone of the letter – demanding a return of the form within 14 days – when one considers the fuss that was made over the IACS forms in 1992. There was one important difference between the two occasions. We were able to send out maps of the farm with the form, whereas farmers had to get their own

CROPPING PROGRAMME

Parish........................

Farm........................

Occupier........................

Date........................

Please specify acreage of each type of Roots to be grown; also whether cereal mixtures are to be cut for hay.

| Field O.S. No. | Field Name | Acreage | Cropping | | | Proposed Cropping for 1946 Harvest | To be undersown in 1946 | | To be cut for Hay in 1946 | For Office use only | Proposed Cropping for 1947 |
			1943	1944	1945		1 year Ley	3 year Ley		Cropping Approved	

Signed........................

Date........................

P.T.O

The following may be regarded as a rough guide for estimating the acreages required for your stock.

Acreage of Crop required per cow, or its equivalent in other stock.

Oats ... = $\frac{1}{2}$ acre } or $\frac{1}{4}$ acre mixed Oats and Beans.

Peas or Beans ... = $\frac{1}{4}$ acre } or $\frac{1}{4}$ acre mixed Oats and Peas.

Roots ... = $\frac{1}{4}$ acre (Kale and Mangolds).

Seeds Hay ... = $\frac{1}{4}$—$\frac{1}{3}$ acre.

Rough method of converting other Stock into the equivalent of cows for feeding purposes.

Dairy Cows and Heifers in calf ... = 1 Cow.

Other Cattle over 1 year ... = $\frac{1}{2}$ Cow.

Other Cattle under 1 year ... = $\frac{1}{4}$ Cow.

Working Horses ... = $1\frac{1}{2}$ Cows.

Ordnance Survey maps in 1992, which was obviously a considerable hurdle. Secondly, it is very significant that we were virtually dictating the way in which people farmed their land. One has a feeling that today's farmers, free of the stress of war, would regard the way we went about it as intolerable interference in the way in which they conduct their business. The third aspect is the way in which we laid down the standards required for the feeding of livestock, and converted them into cropped acres. But here again it has to be remembered that some of the less able farmers definitely needed such guidance in those days, many of them having been brought up on buying in imported feeds from their merchants. The standard of three acres per cow sounds ridiculously high today, but fertilisers were strictly rationed, crop varieties relatively low-yielding, and standards of husbandry not so high, so that inevitably, expected yields from both crops and grass were much lower.

In effect, it was a very good way of educating the B–C grade farmers about livestock feeding. The whole exercise does suggest dictatorship, but the food situation had been very desperate for the past year, with appalling shipping losses, and there was no way of telling at the time for how much longer the war was likely to go on. As it turned out it was to last for another two years, but food shortages continued for many years after that, so that one feels we were justified in taking the actions that we did by the course of subsequent events.

The response to the circular was very satisfactory, with a high proportion of the forms returned within the specified fortnight, and the District Officers had a hectic time sorting them out and preparing the resulting orders for reseeding. The biggest problems in implementing them were encountered in the clay land areas in the south of the county. This was largely because of the difficulty in getting good enough seed beds in the spring to ensure a good 'take' of the seeds. Some farmers held back as they were afraid of losing money because of a crop failure. The technique of direct seeding in the autumn had not been perfected at that time, and most seeding was done either on bare land in the spring, but more commonly by undersowing of oat or barley crops. I favoured seeding under a forage crop at the end of July, but this did not catch on very well, and most farmers continued to undersow.

Alongside this campaign to reseed arable land, we ran a series of trials around the county on the direct reseeding of poor quality grass fields. The first trial had started as early as 1941 in cooperation with Dr William Davies, who was working at the newly established Grassland Research Station in Warwickshire. The need for this had been pointed out by District Committee members who, when visiting farms to get more land for ploughing up for arable production, kept on finding old pasture fields which were not pulling their weight, but which for one reason or another were not suitable for putting into cereals or potatoes – or any other arable crop, for that matter. If these could be got into much better grass by reseeding, then a farmer would feel a lot happier about ploughing out another field for arable crop-ping. We had a number of successful trials and demonstra-tions, but one must also admit that there were quite a few failures – which of course received more attention than the successes.

One of the main problems with direct reseeding was to get adequate consolidation when turning in an old neglected sward, which often had an accumulated mat of semi-decayed fibre in the surface layer. This was largely because of a shortage of lime and phosphate on the pasture, which prevented bacterial breakdown of the residues. Some of these fields had probably never been manured with phosphate, certainly not during the years of the depression. If this type of sward was turned over flat – as it had to be if one was not going to get masses of thistle, buttercup and creeping bent coming up between the seams – it needed very heavy consolidation to prevent the furrow slice lifting and leaving the recently germinated grass and clover seedlings high and dry. This dryness was caused by the coarse mat underneath acting as an insulating layer. If there was a spell without rain, the whole lot dried out and the seedlings died. It wasn't until we began to carry out pre-ploughing surface cultivation to try to break up the mat and incorporate it into some of the surface soil that we began to have rather more success.

By 1944, really big ploughs had come in from America – so-called 'prairie busters' – and these were able to plough some of these rough old fields to a much greater depth, which made them less liable to suffer from drought. Unfortunately they also brought up a great deal of unweathered subsoil, which was difficult to work down to a seed bed, and which could be acid and low in fertility. For the trials, we established the primary importance of building up an adequate level of phosphate, and also of a good dressing of lime.

Farmers were naturally ill-equipped for reseeding in many of those cases where the farm had reverted to grassland in the inter-war years, so the Committee had to be prepared to do a lot of work for them. In 1944, the contract we offered contained a clause which guaranteed that we would do the work again free of charge, if the first attempt failed, though this did not cover any compensation for the loss of grass for perhaps a couple of months while the job was done again. But it did have a good effect in helping to get the reseeding scheme off the ground. The Committee also helped to cover itself by ruling that no field would be included in the direct reseeding scheme if it was capable of carrying a cereal crop.

The farm planning and reseeding scheme went pretty well on the whole and it was continued until the end of the war. Each farmer was sent a new plan every year, with the old cropping plan already filled in, and he then had to insert the plan for the current year and his anticipated programme for the following year. Of course, it was not possible to check physically that every farmer was doing what he said he would do – no 'spy in the sky' planes in those days, since planes were employed on more deadly work! However, the officers had a fairly good idea of what was going on, and the plans were particularly valuable in helping them and the District Committee members to keep the C category farmers up to scratch. It also helped the weaker ones to keep abreast of their work.

These activities came under the aegis of my Cultivations Committee which advised the Executive on policy and put forward recommendations. Running alongside this routine Committee work was Technical Development which became more and more important as the war progressed. It is interesting to look back now at the scope of the programmes that evolved and what we were including in our demonstrations and trials.

Technical Development

From the start, Hugh Mattinson had handed over to me responsibility, under his general supervision, for the early demonstration plots which had been put down on some of the land farmed by the Committee, and also liaison with outside bodies with whom we were beginning to collaborate for specific investigations. As the programme grew, we developed basically two types of work. In the first category were a number of properly replicated experiments which were part of co-ordinated trials across a number of different county WAECs, mostly in the south. To these could be added one or two research type trials of our own and a large number of demonstration trials. These were plots, in some cases replicated two or three times, which were put down at different sites in the county to illustrate to farmers either fresh techniques or new ideas, which might help them in their own farming at home. In this category were obviously variety trials with wheat and also oats and barley, and trials comparing things like dates of sowing and seed rates, or methods of reseeding, such as drilling compared with broadcasting, and direct seeding compared to undersowing.

In the research category, we were collaborating with Rothamsted on two major co-ordinated experiments. These were on phosphatic manuring, and on the use of sewage sludge. The phosphate trial was particularly interesting as it compared two rates of application; broadcast or combine drilled, with a fifth treatment, which used no phosphate at all. The rates used were 1.5, or 3 cwt of superphosphate per acre, and the crop was wheat. The trial was under the aegis of Dr E. M. Crowther, the leading expert on fertiliser usage at the time. We had two of these experiments, one on heavy clay and one on a lighter chalk soil, and they started in 1941 and ran for three years.

Combine drills had only quite recently been introduced, and no one really knew whether placement of fertiliser with the seed at drilling would increase yield, or if it might be possible to use a rather lower rate and thus save scarce fertiliser, if it was placed close to the seed, without sacrificing yield. There were also some national trials using potash, but we did not take part in these.

As might have been expected, there was a significant effect from using phosphate, compared with none at all, but taking the trials across the country as a whole, there was no clear cut result from the combine drilling and broadcasting comparison, which was rather disappointing. Nor, surprisingly, was there a definite result from comparing 1.5 and 3 cwt per acre. The main reason was that with 20 different trials spread over several counties, there was so much variation in the yields over the sites and seasons, that a definite statistically significant result could not be obtained. Our own results also varied a good deal from year to year and between the two sites, but there was never the difference that one might have expected from the different rates of application, on soils that were fairly low in phosphate in the first place. Over the course of the trial there was a slight suggestion that combine drilling gave rather better results, though it was by no means a clear cut effect.

I did, however, learn one very important lesson from the trials, which I have always remembered – that in experimental work one should never rely too much on visual observation. I used to visit the sites fairly often in the early part of the season to score them, and every time the combine drilled crops appeared to establish better, and grow faster for the first few weeks in the spring, and subconsciously one retained the impression that they would be better at harvest. Yet, each time, when the grain was weighed off, there would be very little difference, as the broadcast plots seemed to catch up in the later growth stages of the crop, when it was not visible to the eye. It is extremely easy to jump to conclusions from a visual assessment – especially if one is, perhaps subconsciously, favouring a particular outcome.

We also had another series of trials with phosphate application which were in co-operation with Norman Pizer at Wye College. These were plots investigating various kinds of phosphate such as basic slag, different sorts of ground mineral phosphates, and ordinary superphosphate. The trials were on potatoes and swedes. So far as I recollect we didn't get any very significant differences between the types which was rather surprising, as one would have expected the best results would have come from the superphosphate plots with potatoes, especially. The effectiveness of the mineral rock phosphate did seem to be improved if the soil was acid – as it was on one of our sites. The final report from Wye did show an advantage for superphosphate, contrary to our results, but not a lot of difference between the others. These trials were quite important at the time as phosphate supplies were very restricted, and it was a case of trying to find out whether it was worth using types which would not be used under normal conditions.

The Rothamsted sewage sludge trials were also quite important, as not a lot was known about the effects of sludge at the time, and anything which might boost yields at a time when mineral fertilisers were so scarce, was to be welcomed. They compared two rates of sludge with two rates of farmyard manure, each trial plot also receiving a basic dressing of sulphate of ammonia, basic slag and muriate of potash. One plot received just the straight fertilisers. The sludge and the farmyard manure had to be spread between the ridges, and the fertilisers broadcast on top. I thought that I'd be rather clever and save a lot of time and made arrangements for a gang of Land Girls to be on site to mix up all the straight ingredients beforehand, in order to save having to send a fertiliser distributor over the plots three times. The total area was only about a couple of acres, so there was not a great deal of fertiliser involved.

I got rather hung up in the office on the chosen morning, and didn't get out to the site until about ten o'clock, only to find the girls apparently in floods of tears and threatening revolt.

The reason quickly became apparent for I had forgotten a cardinal point that I used to stress in lecturing at Reading, namely, that when mixing straight fertilisers together you must never mix sulphate of ammonia with a fertiliser containing free lime, such as basic slag. If you do, you immediately get a chemical reaction which produces clouds of ammonia, which had not only made the girls cry, but had also affected their breathing. So the short cut had to be abandoned and we all had to wait about for the fertilisers to be put on individually before the girls could plant the seed potatoes, which was their main job. This little anecdote does serve to emphasise the immense change that took place after the war, with not only the development of a wide range of compound fertilisers, but also the great improvement in their physical condition, and also the reduction in the weight of material that had to be handled. Before the war, quite a number of farmers mixed up straight fertilisers on the barn floor, a most unpleasant job at the best of times, and damned hard work into the bargain, in order to save going over the land several times. Of course, the amounts used per acre were far smaller in those days, as most crop varieties just couldn't stand the heavy dressings used today, but it was a slow, messy and laborious job all the same. Compare that with the mechanical bulk handling of today – farm workers are certainly living in easier times, thank goodness.

It is interesting, though, to see a swing back to the use of single straight fertilisers again, as the soil reserves of phosphate and potash have risen to such high levels on many arable farms after some 40 years of higher applications, that complete fertilisers are not always required.

The need to restrict the timing of nitrogen applications for environmental reasons also indirectly militates against the use of compounds.

This illustrates the point that if you live long enough in farming, everything goes round in cycles, and practices which fall out of fashion for various reasons, come back into use again, though often in a rather different form, due to some new technological development which makes them feasible once more. Tower silos are an excellent case in point.

Quite a number of them were put up in the inter-war years – huge concrete edifices, requiring a large labour and energy input. These were dynamited out of existence in the '50s as labour got scarcer, and pits and clamps became fashionable. Then along came the brief era of the steel towers that sprang up in some of the dairying districts, but it wasn't long before they, too, turned out to be costly in their energy requirements, and many fell back into disuse once more. I must say that I could never really see the point, even before the war, when we had a big concrete brute on the Reading farm, of spending an awful lot of energy chopping up grass so that you could then blow it up to vast heights, and then had to use a lot more energy to get it down again, even if the product might be a bit better than anything else that was available.

But, back to wartime, and experimental work. The grassland trials carried out in co-operation with the Grassland Research Station were on private farms with different seed mixtures on two contrasting soil types – heavy and light. Farmers had plenty of opportunity to inspect the sites on farm walks. We also did seeds mixture on other farms, mainly to demonstrate the value of cocksfoot on very light soils.

Another co-operative association with an outside scientific institution came later in the war, and this was with Imperial College in London in their experiments with new herbicides, and this was destined to influence the future course of my working life.

A Demonstration Farm

It must have been in the summer of 1942 that I was asked to write a report on a sizeable farm in the central south of the county, which had fallen vacant through the death of the tenant. It was on an estate which owned several farms in the locality, but, unfortunately, the owner, an officer in the RAF, had been reported missing on active service, and there was no evidence as to whether he was still alive. The trustees and their agents were not anxious to tie up the farm on a long term tenancy, until there was more definite news of the owner, and suggested that the Committee might take it over and farm it for the duration of the war. It was over 300 acres, with good sized fields, but on a rather silty weald clay which was difficult to work in a wet year. It was not as heavily wooded as many farms in that area but was in a pretty dreadful state. The arable fields were full of those weeds which are the curse of wet clay farms: creeping bent, creeping buttercup and thistle, and catmint, not to mention annuals such as white and yellow charlock, speedwell, and poppies. The pastures were no better: agrostis dominant, and they walked sodden to the tread. That was not surprising because the ditches were silted up, and hedges were encroaching into a number of the fields. To make matters worse, the house had been empty for quite a long time, and had been requisitioned by the Army. It was in a very dilapidated condition as were the fairly extensive range of buildings, with the exception of a large Dutch barn, which must have been put up by the estate not long before the war. The agents were probably rubbing their hands with glee at the prospect of someone taking it over lock stock and barrel as it stood. I doubt if they could have found a good tenant to take it on in that condition, even if it was offered rent free for a year or two.

I have perhaps painted a rather depressing picture of the place, but it was not untypical of many of the smaller farms in that area at the start of the war as a result of the previous 50 years or so of depression. Many of them had drifted into milk production as a last resort, as the price of cereals made arable farming quite uneconomic on this wet difficult land. Fields had been allowed to tumble back to very poor quality grass, and the farms became little more than ranched grassland for the stock. Labour was cut to a minimum, and was insufficient to keep the ditches in proper repair and the hedges trimmed back. Quite a lot of the land had been tile drained in the drainage booms of the mid 19th century, but once the outfalls in the ditches got covered up, the whole system became useless. It was really a vicious circle of low productivity reducing incomes, which led to more cutting of corners, and still lower productivity.

When the war came, a lot of these farmers were very

ill-equipped to deal with the demands of the ploughing up campaign, and a number of them fell by the wayside. However, those that did survive profited from the higher prices and managed to pull their farms round, often with the help of better-equipped neighbours, or through the use of the Committee's machinery and labour services.

I had been thinking for some time that it would be an advantage if the Committee could have a whole farm under its own control which it could use as a focal point for demonstrations and trials to help us in our technical development and advisory role. Having walked this farm, I could see that it had a definite potential for this, if we could pull it round, in spite of its obvious disadvantages of having no proper house or satisfactory buildings. So I put in a strong report for it to be taken over, which in due course it was, and we were able to go in and set about its reclamation. There was little alternative but to fallow quite a high proportion of the arable land in the first year, while the hedges were cut back, and the ditches dug out (mostly by Italian prisoner of war labour which was just becoming available), and the land mole drained. When the ditches were cleared, most of the arable fields turned out to have been tile drained, probably in the 1850s–70s, and with rodding out, some of them started working again. One of the great pleasures of farming a heavy land farm is to uncover an old tile drainage system, and to see the pipes discharging drainage water once more, or to spend money on a new system, and to watch the water gushing out after heavy rain.

It took 12 months hard work to get the farm working again, and we were able to patch up the buildings well enough to keep bullocks and get straw trodden down into farmyard manure to help to rebuild fertility and soil structure on some of the land. But it was really two years before the farm began to be reasonably productive, and we could put down demonstration plots. By the summer of 1944, we had established cereal variety trials and demonstration plots on methods of seeding arable land back to grass, such as drilling versus broadcasting, and sowing direct versus undersowing. This was topical in view of the reseeding programme which by then had got under way in the county. There were also seeds mixture trials, and a field comparison of methods of preparing land coming out of a one year ley for wheat. These were a bastard fallow, ploughing after a second cut of seeds hay, or ploughing in the second growth. The bastard fallow gave the best results here, largely, I think, because we got a much better seed bed for the wheat. Then there were plots with mangolds comparing salt with potash combined with different rates of other fertilisers, and trials on the effect of varying levels of phosphate application when reseeding.

In the early stages, we used the farm for occasional evening walks, but by 1944 it was well enough established to give us the confidence to mount a major field day, at which, apart from tours of the plots, we had competitions for Land Girls, demonstrations of drainage and hedge laying, and so on. It was a very successful day, with several hundred farmers turning up. We mounted another one the following year on a private farm at the other end of the county, but this was more livestock based, with Winston Rowe playing a major part in its organisation.

In addition to these major central operations, the District Committees also organised farm walks on a local scale, which were popular, providing farmers with the chance of seeing farms that otherwise they might never see, and an opportunity for putting across new ideas. In addition, we fostered discussion groups in the winter months, brains trusts, and quiz competitions. Similar clubs formed in those years are still active today in many counties. It was a busy time getting round to as many farm walks in the summer, and discussion groups in the winter as one could manage.

Following Hugh Mattinson's death in 1943, there had been a gap of some four months before his successor, Tom Creyke, arrived. He was only a few years older than I was and a tough Yorkshireman who had previously been Deputy Executive Officer in Leicestershire. He had gained useful experience there in turning what was almost an all grass county in 1939 into one of mixed farming and had faced many of the same sort of problems that we had. As in my Sudan experience six years previously when Edward John had been off duty for six months, I found it a bit difficult relinquishing the power that I had assumed. However, Tom handled the situation very tactfully, and we got on well together. I remember him putting me in my place pretty firmly once or twice, which was fair enough, as he *was* my boss, but I hope that I played the part of the good staff officer, and took it all in good part.

He was a complete contrast to Hugh, particularly in pushing things along. He took quick decisions, whereas Hugh tended to sit on things for rather too long before making up his mind, and sometimes one didn't quite know where one was with him. Much more of a businessman type, he was sharp and to the point, and capable of being ruthless, if the occasion required it, and his later career as an officer in the World Bank probably illustrates that point very clearly. He quickly appreciated that I was in a somewhat anomalous position without any real official status and got the Executive to agree that I should be officially designated Chief Technical Officer, with responsibility for all advisory and technical matters and for the farming of the Committee land. At the same time Winston Rowe was promoted to Chief Livestock Officer with responsibility for development work in that sphere, especially for dairying development.

The second very positive action that he took on arrival was to get us more assistance, realising that both Winston and I had really rather more on our plate than we could manage properly. It was then just coming up to the end of the summer university term, so he got in touch with Reading to see if they had any final degree students they could recommend. Some universities with strong agricultural departments had been allowed to maintain their agricultural courses, as the Government realised that it was essential to keep a flow of young graduates coming in to take the place of those who retired or became otherwise incapable. We fortunately managed to collect four young graduates, one of whom came to me, one went to Winston, and two were appointed to help two of the hardest pressed District Officers.

I was very lucky indeed to get a most delightful assistant, and I could not have asked for a better one. Johnny Gilmour was one of the most cheerful and easygoing people that I have ever come across and worked with. Nothing was too much trouble for him, and he was happy to work all hours, and would come up with a cheery grin at the end of the longest day out in the field working on trials and experiments, or getting home late after a farmers' evening meeting. I can only hope that he found working with me as easy as I found it was to work with him. I only met him once after the war, so get in touch, Johnny, if you happen to read this.

The Birth of Artificial Insemination

I think that it must have been about the middle of 1942 that we first got wind of a proposal by the Ministry to set up two AI stations, at Cambridge and at Reading, and that the Reading centre was likely to have two sub-centres attached to it, one of which might be at Guildford. The object would be to spread the catchment area as wide as possible, and to get enough customers in the early stages to make it a viable proposition. Winston Rowe was very excited about this, as was the Executive, which, as I have said, was a very progressive and forward looking group of farmers. The members appreciated how valuable such a service might be for the large number of quite small dairy farmers in the south of the county, few of whom had much chance of buying a decent quality bull. As a lot of them had been producer retailers – and still were – a good deal of crossbreeding had gone on. Many of the smaller herds presented a pretty motley appearance, with traces of Jerseys, introduced to improve the cream line in the bottle, or Ayrshires to step up yield, or even a little black and white, since Friesians were beginning to get rather more popular by then. Most of them when they needed a new bull went to a local breeder who had a good reputation, and bought a bull calf out of one of the better cows in the herd, and hoped that it would improve the yields of the dam's present daughters, which, of course, it seldom did because of the vast genetic variation in most breeds at the time.

The AI centre was duly established at Reading and our sub-centre just outside Guildford, in 1943, and it then became a matter of trying to sell the service to farmers. One would have thought that a small farmer would have jumped at the chance of saving himself the trouble, cost, and danger of keeping a bull simply to service a small number of cows. But in the early stages, it was very much the larger and more progressive farmers who took up the service. I suppose though that it was natural for a farmer who had remained a small scale operator, probably doing the milking himself and seldom getting off the farm, to be distrustful of such a revolutionary idea. He would worry that the system might not work and that he would be left with cows not in calf, and a reduced income. Few really knew what it actually cost to keep a bull for just a handful of cows, and thought that AI would be more expensive. In fact, the lower cost of AI was one of the points that Winston stressed in trying to sell the system in his advisory work. AI caught on quite fast with the better farmers, and infiltrated down from them to the smaller ones, once they could see that the system did work as well as the bull did – and sometimes a great deal better, if they had infertility from venereal diseases in their herds. This was particularly common in those herds where small scale farmers shared a bull with several neighbours, which was still quite a common practice at the time.

It took some years for the number of bulls recorded in the national statistics to decline significantly, as some farmers continued to adopt a belt and braces approach. But when more independent cattle breeding societies were established in different parts of the country, following the success of the Ministry's pilot scheme, they began to fall significantly in the late '40s and early '50s. I was naturally delighted in view of my experience of the system at Cambridge, to see it expand so successfully in farming practice.

Another Technical Breakthrough

Both 1943 and 1944 turned out to be milestone years for technical development in farming, the former because of the introduction of AI and then in the following year we first began to get reliable information about a completely new concept of weed control. In that year, there is just a brief note in my report to our Technical Development Committee under a heading 'Spraying Trials', which ran, 'These are being carried out by a team under Dr Geoffrey Blackman at Court Farm, Hambledon on White Charlock and Spurrey and the control on some plots seems to be very successful.'

Earlier that year, Blackman, whose research unit was then based on Imperial College in South Kensington, had approached us, as being conveniently close to London, to seek our co-operation in providing sites where there were problems with specific weeds, on which he could assess the effects of the newly discovered systemic hormone weed-killers (the word 'herbicide' did not come into common use until a few years later). The two that he was using were MCPA and 2.4.D, which achieved their effect on specific weed species, by distorting the growth pattern of the plant to such an extent that it could no longer survive. Up until that time, chemical weed control had depended on a blanket scorching effect on the herbage of the plant by a chemical — which was generally sodium chlorate. But that could only be used on weeds on bare ground, as it killed the crop as well. The great advantage of the hormone compounds was that they killed a range of dicotyledonous broad-leaved weeds in a monocotyledoneous narrow-leaved crop such as a cereal, i.e. they were selective. This new development sounded very exciting, and we jumped at the chance of being able to assess its likely effects at first hand on our doorstep.

Over the next two years, we managed to find a lot of sites for the use not only of the hormone sprays, but also for experiments with DNOC (dinitro ortho cresol). This was a yellow dye, which it had been discovered had herbicidal properties. Though it was essentially a contact herbicide, it did have some selective effects on broad-leaved weeds in cereal crops. It was not a very pleasant material to handle, which was one of its disadvantages, but it also affected some weeds which were not susceptible to damage from the hormone sprays.

Blackman had a strong team of young scientists working for him and they did all the field work, while it was our function to find sites on private farms or on our own land to meet his requirements. This was easy as farmers were usually very keen to co-operate if they thought they could learn something to help them control weeds, other than by cultivation. I cannot claim to have made any personal contribution to the revolution in weed control methods that followed the introduction of systemic herbicides, but, as with the advent of AI, and the use of pneumatics on farm vehicles, I count myself lucky to have seen at first hand the initial work that led to the introduction of techniques which so revolutionised farming practice.

Mention of DNOC brings to mind an amusing episode. In June 1945 I was rung up one day by a very progressive farmer near Reigate who was always keen on trying anything new. On this occasion, he had put down about eight acres of Timothy grass for seed purposes to harvest in the following year. (There was a strong demand for grass seed at the time due to the amount of arable being put back to grass, and there was good money in growing the seed.)

In this case, he had sown it in rows about 14 inches apart, so as to be able to hoe between the rows, and had got a good 'take'; but the weeds had come up in their millions within the rows, and the Timothy being a rather slow starter was being completely smothered. When I went to take a look at it, I found almost every weed in the book. However, they were mostly annuals, which if destroyed would leave the crop relatively free the next year. Some of them I already knew by then would be controlled by 2.4.D, but others I thought would be resistant. It was clear that quite drastic action would be needed if he was to save the crop, which had already cost him a lot of money as the seed had been very expensive to buy in the first place.

I decided that with the thick cover of weeds, DNOC would probably not do the job very efficiently as it would not reach all the leaves, and in addition I didn't know whether the grass might be liable to damage from it. Equally, I realised that there were quite a few weed species there which would not be touched by 2.4.D or MCPA. So I rang up Geoffrey Blackman to ask if he had ever mixed DNOC and 2.4.D together, and whether the Timothy might suffer scorch from the DNOC. He said that, as far as he knew, nobody had ever mixed them together before, but he did not think that there was any likelihood of a chemical reaction between them, or that the DNOC was likely to damage the Timothy. If we decided to give it a go, he would be most interested to hear how we got on. The farmer concerned was always one willing to take a risk, and could not bear to think of the money he had already spent going down the drain. So we went ahead and with Blackman's assistance got the chemicals which were not yet on the market.

Then one lovely sunny Saturday morning with the larks singing above us, Johnny Gilmour and I met the farmer on the field, with his tractor driver and sprayer. In fear and trembling, with visions of weeping Land Girls very much in mind from the previous mixing episode, we poured the chemicals in the appropriate quantities into the tank. Geoffrey Blackman was right, there was no vast explosion, nor any sign of a chemical reaction, so we had cleared the first hurdle successfully. Sprayers were not terribly reliable then, and there was always the risk of blocked nozzles, so Johnny and I walked behind to make quite sure that everything was all right. When we broke for lunch, we realised that we were already pretty yellow from the DNOC, but as we were only half through by then, we decided to carry on as we were. The job was finished at about 4 o'clock, when the farmer said, 'Now, you'd better come in for a bath, and we'll have a cup of tea, and then I've got to be in Brighton for an appointment at 6 o'clock. Why don't you boys come down for a drive and some sea air, to keep me company.' I looked at Johnny, who said he had got nothing on that evening, and I was also free; but I said to the farmer that I didn't think we could go down there with yellow shirts. 'Oh, that's all right, I'll lend you a couple of shirts, if you'll let me have them back later.' It was a beautiful evening, so we said we'd be very happy to go.

But what we had not realised was that DNOC was very much a dyestuff, and didn't come off one's skin at all willingly, and that our hair and eyebrows were also tinted bright yellow. Scrub as hard as we could in the bath, the damned stuff just would not come off, as, in fact, it didn't for quite a few days.

I suppose in the liberated times in which we live today, two young men walking around Brighton on a Saturday evening with artificially bronzed faces, and yellow tinted hair, eyebrows and finger nails, probably would not merit more than a second glance. But I was very conscious of some strange looks in the pub we went into for a drink, and the restaurant where we had a meal, even though we were not actually asked to leave the premises. Perhaps if it had been in Hove and not Brighton, we might have been!

I suppose that we should consider ourselves lucky to be still alive today in view of the fact that DNOC has now been banned as a herbicide because it is considered a serious risk to human health. It is perhaps also a not very praiseworthy illustration of the slap happy attitude that most people had in their approach to handling chemicals when they first became more widely used in farming. It was to be a good many years before we began to be pressurised to wear protective clothing. To go back to the field of Timothy – we did manage to get quite a good kill of weeds from the first spraying, but it required a further application of DNOC later in the summer to kill off those that had regenerated.

Our co-operation with Geoffrey Blackman continued to provide good information about the new herbicides, but we could not use this in advisory work, as they were not yet on the market, and Geoffrey had to pull strings if we needed any for research or demonstration purposes.

A RETURN TO EDUCATION

It was in 1944 that I began to get involved once more in agricultural education in a minor way. The year before, some local farmers' sons had got together and formed a Young Farmers' Club in Guildford, partly at Winston Rowe's instigation. Some of them felt that they wanted rather more technical information than could be obtained from the normal club activities, so they suggested that we might start up some evening classes at the local technical college. Winston and I both thought that this would be a good idea, since at the time there were few Farm Institute courses available, and most of the young people involved could not leave their jobs for a residential course because of the shortage of labour.

The college was quite happy to provide facilities, so in the winter of 1944, Winston and I ran a series of classes. He covered livestock, and I concentrated on soils, crops and related topics. We had, I suppose, about a dozen very keen students and found it a rewarding exercise because of their enthusiasm. The following winter, we put on another course in the east of the county at Redhill but this did not recruit so well, and we only ran it for a year. It was a great pleasure incidentally to be invited back to the 50th anniversary celebrations of the Guildford YFC in 1993, to see many of the founders once more and also to meet again the mother of one of them, now well into her nineties, whose sumptuous farmhouse teas are still a happy memory.

At the same time that this was happening, the Ministries of Agriculture and Education were taking steps to implement some of the findings of the Luxmoore Commission on Agricultural Education which had reported in 1943, and recommended amongst other things an increase in the number of residential Farm Institutes in the country. This Commission had been prompted by a very creditable recognition that farming was quite rapidly moving into an era of technological progress, and that the provision of technical education for those in the industry had been seriously neglected. This had been shown by the unpreparedness of many farmers and their workers for the demands made on them by the food production campaign in the early years of the war. It was known that only a tiny proportion of the labour force had had any formal training in agriculture at all.

Several of the leading farmers in the county, both on the Executive and the District Committees were very keen that Surrey should have its own Farm Institute, and not have to rely on Sparsholt in Hampshire, or Oaklands in Hertfordshire for its training. After a great deal of discussion, it was decided to establish one under the aegis of the county council, and the hunt was on for a suitable site. I was not personally involved in this, as it was Tom Creyke's job. There were quite a few large properties available, as the troops had by now moved out to France, and some big houses which had been requisitioned were empty, but most of them did not have a farm attached. The choice finally fell on Merrist Wood, a large, quite modern house with a farm, at Worplesdon, about five miles north-west of Guildford. It wasn't ideally suitable, either for its location, which was too far west, or for the quality of its land, which was very light and not very typical of the main agricultural soils in the county. However, it seemed to be the best that was available at the time, so it was purchased by the county council, in the face of a good deal of criticism.

In fact, as things have turned out 50 years on, it may not have been such a bad choice after all, for with the dramatic decline in the labour force in straight agriculture, and the changed functions of the former Farm Institutes into those of colleges catering for all types of countryside activities, the site has definite advantages. Surrey was always a county with a strong horticultural bias, both in vegetable production to meet the needs of the Greater London market, and also amenity horticulture for supplying the huge suburban population with garden plants of all types. The horticultural industry has expanded enormously as the standard of living has improved since the war, and Merrist Wood is very well suited to this type of production. It is also very well situated for covering all aspects of environmental education, and for dealing with the problems that arise at the interface between large urban populations, and the open countryside. So, as the situation has evolved over the past 30 years especially, the purely agricultural role of the college has declined very significantly, while the horticultural and environmental role has expanded greatly in its place.

The college opened its doors to students in the autumn of 1945 with Ted Hankinson, always known as Hank, who had been Vice Principal at the Cannington Institute in Somerset, as its first Principal. But he had not had time to collect many members of staff, except for two young graduates, one of whom, John Pollard, went on to become the very highly regarded Principal of the Berkshire College at Maindenhead. So for the first year, Winston Rowe and I were pressganged into taking on the main lectures in animal and crop husbandry respectively. Fortunately, I had kept my old Reading lecture notes, so did not have to spend much time in compiling a new set, as I could just embellish the old ones with the experiences gained in looking after all the Committee land for the previous five years, and incorporate all the technical advances which were taking place. We each did three one-hour teaching periods a week, and I enjoyed getting back to teaching again, as the first intake was a very nice, keen bunch of about a dozen students. By the next autumn, Hank had been able to build up his staff, but I had by then already left Guildford, so would not have been available in any case. But it was enjoyable to go back to the college from time to time in later years as an examiner for the National Certificate in Agriculture, and to see how it had expanded from its quite humble beginnings.

THE APPROACH OF PEACE

It was still a very busy time, for as the war in Europe drew to its close, some of the owners of the land that we had requisitioned over the previous five or six years began to look forward to its future use, and in the case of the house building companies and developers to ask for it to be handed back. This could only be done by the Minister of Agriculture on the advice of his local Land Commissioner. So in late 1945 and the spring of 1946 I found that I was spending a lot more time again with Tom Sutton. Only this time, the question was whether the land could be given up, rather than taken over. There were two main factors that we had to bear in mind when coming to a decision. Firstly, there was still a severe shortage of food and very strict rationing, and though the threat from U-boats had gone, shortage of shipping and also of foreign exchange limited the amount of food that could be imported.

This situation seemed likely to persist for quite some time – as indeed it did, for several items did not come off the ration for a further five or six years. It was, therefore, imperative that we should continue to grow as much food at home as possible.

The second consideration was one of planning. By the start of the Second World War, there was already very considerable disquiet being voiced at the lack of adequate planning controls on building, which had allowed speculative builders to run up cheap, shoddy houses, bungalows and ugly industrial buildings virtually where they pleased. This had led to dreadful examples of ribbon development in many parts of the country, and unnecessary spoliation of the countryside – examples of which we can still see only too often today. The Government accordingly was already planning new restrictive legislation to stop this happening again in the future, but it had not yet got a Town and Country Planning Act prepared to put before the House. So until that happened, and also until essential building materials were in more plentiful supply, the instructions were to go easy on releasing large areas of land for building.

Of course, Tom and I were really only entitled to give advice as to whether the land was still vitally important for food production, as we had no brief to advise on straight planning issues. However, at the same time we were told to take into consideration the quality of the land and whether it would be a long term national asset for it to remain in food production. We were fortunately not involved with a particularly hot potato across the county boundary to the north of us, which involved this very issue. That was whether high quality market garden land which had been requisitioned for the use of the RAF should be retained and further expanded for the development of a national airport for London, i.e. Heathrow. I well remember that there were many arguments on this issue. As far as we were concerned at the time, we generally found it difficult to make any strong cases for retention, since much of the land on the North Downs on the southern fringe of the suburbs was intrinsically Grade 3 land or worse. It was already generally divorced from any whole farming units, by building which had already taken place just before the war, and it clearly had no future role to play in mainstream farming. Its productivity was low, so that the state would not lose much food, if it was built over. Occasionally, there was a borderline case where future planning controls would clearly be involved, and Tom Sutton would tactfully recommend a delay in relinquishing the land. Even though we did not have a great deal of difficulty in coming to our decisions, it was an interesting exercise, dealing with issues not previously encountered, and Tom's land agency approach provided me with excellent experience.

Looking back, I certainly cannot complain about a lack of variety in the type of work with which I was involved for the Executive Committee, nor at the scope provided to develop that work in its different aspects. I was given wonderful opportunities for gaining experience in all sorts of fields, and the support given by members of the committees both collectively and individually was tremendous, and is something for which I shall always be most grateful. The opportunity of being involved in the technical changes that were taking place in agriculture, both through close contact with the most progressive farmers in the county, and also through the medium of the land that we were farming ourselves was invaluable. Also of great use were the contacts with the research centres, such as Rothamsted, Wye College, the Grassland Research Station and Imperial College which kept one abreast of progressive thinking. It was an exciting and stimulating time, by the end of which I could really feel that the years of apprenticeship and first practice were over, and that I was now at the threshold of maturity. This feeling was supported by promotion to the rank of Deputy Chief Executive Officer towards the end of 1945. I don't seem to remember that it meant any increase in salary, or that it made much difference to the range of work that I was doing! Fortunately it did not mean, as I had feared it might, more inside administration which would have been the last thing I wanted.

LIFE IN WARTIME

As might be deduced from this account of the Committee's work, and my part in it over the six years of the war, there was not much time for other outside activities. We worked long hours, and as the war progressed, there were frequent evening meetings to attend – discussion groups in the winter and farm walks in the summer. Petrol was severely rationed for private purposes, which restricted social events, but coupons were available for official work, so I had enough to get round the county for overseeing our farming activities and the advisory and development work. In addition, I had a small allocation for Home Guard duties as my unit was based at the other side of the town from where I was living. So I was not restricted for movement within the county, but did not have enough for going off to Cornwall for a holiday more than about once a year.

I had joined the Home Guard within a few weeks of starting the job, and though I had got over the worst of the symptoms of the Sudanese collapse by that time, I was still not really very fit, and found the Home Guard fairly exhausting if any major exercise was involved. I had been very lucky in finding some excellent digs where I was very well cared for, and there was so much to do in getting to grips with the job that there wasn't time to worry about one's health. Soon after I arrived, a young Wye College graduate, Adrian Pelly, joined us for about nine months while awaiting call-up for the Navy, and we became friends – and remain so to this day. He was living with a cousin, married to a London stockbroker, who had a lovely house in the country some six miles out of Guildford. They were a wonderfully kind couple, and it became almost routine to go over there on Sunday afternoons after the morning Home Guard parade. This continued after he left for the Navy in the autumn of 1941, and their house became almost a home from home.

The Home Guard was an amazingly democratic organisation drawn from every conceivable strata of society. Our company commander was an ex-Warrant Officer retired from the Army who worked for us in the WAEC as the manager of one of the machinery depots, and our platoon commander was a City stockbroker. Our ages ranged from 16-year-olds awaiting call-up to 60-year-old veterans of the First World War. There was a good measure of others in between who were in reserved occupations of one sort or another, or who were medically unfit for active service. The weekly all-night duties and two hour patrols enabled one to get to know most of the members of the platoon very well and to get an insight into their diverse worlds.

Having been through the OTC at school, followed by the Khartoum training course, I was fairly soon given a couple of stripes, and then in 1943, became a full blown sergeant, which did not mean very much, except for being in charge of the night guard once a week. This involved a stint from 8.00 p.m. to 6.00 a.m. in the guardroom which was a large garage of a private house in the wealthier residential area to the south-east of the town. We had a biggish area to cover, including part of Shalford village, the Chantry Hills and then up to the top of the chalk downs behind the town. Scenically, it was a beautiful area by daylight, but that naturally did not concern us on our nocturnal perambulations.

Exactly what our function was on these patrols is rather obscure. In that area, there was not the remotest possibility of enemy troops sneaking in by night in view of the very wooded nature of the terrain. Nor was there likely to be any very sinister fifth column activity, for the same reason. We were supposed to keep a close look out for anyone acting suspiciously. We also probably played a civilian role in helping to deter burglars in the vicinity, and a public relations role, in that many people *were* nervous about parachutists and so on in the earlier days of the war, and slept more soundly in their beds in the knowledge that there was someone keeping a close watch on things outside. I only ever remember one incident of note over five years of patrolling in the small hours, and that was in 1944 soon after D-Day. At about four o'clock in the morning we were on the top of the downs and heard a plane coming in from the east making a most peculiar noise, rather as if it was flying on only one faulty engine. We couldn't see anything, but it came almost over us flying fairly low, and we assumed that it must be a returning bomber that had been shot up, trying to limp home to base. About five minutes later there was a flash in the sky to the west, and the sound of a dull explosion. We said something like, 'That's another one gone, let's hope that the crew got out in time', and I made a note of it in my report for the night. It wasn't until two or three days later that we heard that the first of the long expected flying bombs, the V-1s, had come over, and that some of them had gone very much off course from their London target – as this one had obviously done. Fortunately it had landed in open country, and no damage was done. We had been led to expect something of the kind for quite a long time since we had been involved in finding additional sites in the east of the county for the barrage balloon defences of London, which were clearly not designed to deter conventional high flying bombers. Nothing had been said officially, in order to avoid alarming the public, and we had rather forgotten about it by the time the first ones came over.

After that we had a number of V-1s over us, usually on windy days, and I well remember one Wednesday morning at the regular meeting of the full Executive Committee when the sirens went, and we heard in the distance the now familiar, but still ominous *phut, phut, phutting* of the V-1 engine coming closer and closer. Suddenly, the members seated round the large oval table, all with one accord pushed back their chairs and dived down underneath it, and we came face to face with our opposite numbers across the table legs, and we stayed down till the noise of the engine had passed.

It took me back most vividly to my early childhood days in the First World War, when on a number of occasions my mother woke us hurriedly in the middle of the night, and put us under our very large and solid oak dining-room table. The reason on those occasions was a Zeppelin passing over the house on its way to making raids on London. We lived then on the extreme northern edge of Epping Forest, which runs from the small town of Epping in Essex down to the northern fringe of London.

The forest was apparently wide enough to provide them

with a relatively safe corridor by which they could approach the capital without too much risk of being downed by anti-aircraft fire. It cannot have been entirely safe, however, since one was shot down only a few miles away from our house. I can still hear in my mind the dull throbbing sound of the Zeppelin's engines as we crouched under the table. I clearly remember one morning in 1917, when I was playing in the garden, and became aware of approaching aeroplane engines, and one of my sisters rushed out of the house, and dragged me under that same dining-room table. These were the German planes on their way to making the first daylight raid on London, and I vividly recall the shape of planes in the sky, as I was taken into the house. I believe that it proved a very costly raid and they did not come over us again in the daytime.

Of course, the Committee needn't really have panicked that morning, as it was already known that there was virtually no danger from a V-1 rocket so long as one could hear the engine running, since it would remain airborne. It was when the sound of the engine stopped, that one had to dive for cover, since once it had cut out, it was bound to hit the ground within a minute or two. It is all very well to say that we need not have taken such sudden action, but the noise made by those rockets was somehow singularly scary and sinister, and created something of a feeling of panic as it got closer, and closer, and closer, and one dreaded the moment when it might suddenly go silent, as the engine cut.

But even they were preferable to the soundless V-2 rockets which came later, and which struck out of the blue, and with a much more powerful warhead. I shall never forget one lovely clear spring afternoon in 1945, when making an inspection of land on a hill near Banstead, looking down over London spread out beneath us. Suddenly, the sirens went, and almost immediately there was a huge eruption of dust and smoke from an area somewhere just south of the river, and then seconds later the distant sound of a heavy explosion. It was a V-2 bomb, which had brought sudden death, terror, and destruction out of the clear, blue sky of a peaceful and placid afternoon. I appreciated then, more than at any other time, the utter wastefulness, the stupidity, and the horror of an all out war, which kills and maims completely indiscriminately. If I had been living in London, or been in the Services, I would have appreciated it long before. It was that grim contrast between a beautiful spring day when the birds were singing and everything was coming back to life again after the winter, and the sudden senseless termination of life, that really brought it home to me.

I suppose that there was rather an element of *Dad's Army* about the Home Guard, but we did take it all very seriously. If there had been an invasion, I don't imagine for one moment that we would have been able to hold up crack airborne troops for more than about five minutes. Apart from our own personal old First World War Lee Enfield rifles, we had precious few machine-guns and mortars available, and were not fully trained to use them anyhow. But we could possibly have played a useful supporting semi-civilian role behind the lines in an emergency, guarding important installations and keeping supply lines open. We certainly wouldn't have stood a chance as combatants against front line troops in spite of our four hour training sessions on Sunday mornings, and occasional field day exercises.

It was in connection with the Home Guard that I had my first encounter with the police in the winter of 1944 after a Friday night battalion dance at the Angel Hotel in the centre of the town. In the office the next morning, the phone bell rang. 'This is the Guildford Police speaking. Your car was recorded as being at the Angel Hotel last night. We would like to know what you were doing there and the source of your petrol.' It so happened that I had been the Duty Sergeant that evening, and had driven the Duty Officer, our company commander who as I have said, worked for the WAEC, round to the two guardrooms in different parts of the town. I told the police this, and said that the second guard room which we had visited was on a direct route between the Angel and where I lived, so we had dropped off at the dance on the way home. The policeman said he'd come round and take a statement. It wasn't till about ten minutes later that I suddenly realised that, in fact, we hadn't visited the two guardrooms in the order that I had stated, but the other way round, and that the hotel was nowhere near on a direct route between that guardroom and the Mattinson's house where I was then living. I hastily rang the Machinery depot to try to warn the Duty Officer to ask him to perjure himself and corroborate my original story. Of course, he was out, so I left a message to ring me urgently as soon as he returned. The policeman arrived to take my statement, and of course the phone then rang, with him only about two feet away behind the desk. Naturally it was the Duty Officer, and I had a moment of panic, thinking that the policeman must have heard the name on the phone. I rapidly collected my wits, saying, 'I'm terribly sorry, but I'm rather tied up at the moment, can I ring you back in a few minutes, as I would like to have a word with you.' Oh, what a tangled web we weave when first we practice to deceive – I never realised till then how perceptive was that childhood rhyme. I rang him back as soon as the copper had left, and he agreed to support me. They never contacted him for a corroborative statement, so my near panic and feelings of guilt were quite unnecessary. But it proved that one always had to be very careful when using a car to have a good reason for being in a particular place at a particular time.

From a personal point of view the years 1942 and 1943 were significant for various reasons. Late in 1941, my mother was diagnosed as having cancer, and was given some three months to live, which turned out to be an accurate forecast, for she died in the following spring. With wartime travel being difficult and petrol not available I was not able to get to Cornwall as often as I would have liked to support my father and sister, though they did have the help of one of our aunts for the last weeks. The loss of a parent in a closely knit family must always be a rather traumatic experience, and I missed my mother very keenly.

Then in January 1943, I went to the wedding of one of our District Officers and that night started being sick with a very nasty abdominal pain, which by morning was fairly obviously appendicitis. Fortunately, they operated that afternoon before it ruptured – something that had nearly cost the life of my younger sister 20 years before. After four days in hospital, I was moved to a convalescent home, manned almost entirely by VAD's, where we had a whale of a time for another ten days, so much so that I was quite sorry to go back to work.

Even after the appendix, I was still not feeling right, and the doctor referred me back to the surgeon who had done the operation. He had a local reputation for being a gall bladder fanatic, and after listening to the Sudan story, he poked a finger into the top of my abdomen, whereupon I shouted, 'Ouch', in a loud voice, as it hurt quite a lot. 'Right,' he said, 'you've got a gall bladder infection, I'll take it out for you, if you like.' In those days, it wasn't the commonplace operation that it is today and nobody seemed to know if there were likely to be any longer term consequences. However, with my agricultural training, I knew that horses, and some other species got along quite well without a gall bladder so I didn't really see why I shouldn't; and if it meant that I might be fully fit again after some four years of feeling below par, it would certainly be worth the risk. So I said, 'OK, go ahead,' and early in August, I had three pleasant weeks in hospital again, visited at frequent intervals by a succession of delightful girls from the office, who all seemed suddenly most concerned about my welfare (none of them, I fear, succeeded subsequently in getting me to the altar!). I felt a lot better after that operation, but it was another four years before I felt really fit after various dietary changes.

I had hardly got back into the full swing of work again, before we had the awful experience of Hugh Mattinson's sudden death, and it was fortunate that I was feeling better by then, and ready to take on the extra work and responsibility that resulted from it. As soon as his estate was settled up, it became clear that Elisabeth, his widow, had been left in very straitened circumstances, with three teenage children, all at quite expensive stages of their lives. She decided that she would have to take in two lodgers and go back to part-time secretarial work to make ends meet. So I broached the possibility that I might give up my digs (my landlady was anyhow thinking of moving at the time), and come and live with them – a proposal that she welcomed (metaphorically) with open arms. So, in the spring of 1944, I moved in, becoming almost part of the family, and spent the next two and a half years very happily there. The girls by then were about 18 and 16 years of age, so we began to lead quite a social life, with occasional concerts and dances which became more frequent as the war pressures steadily relaxed. Once again, I had really fallen on my feet in finding a home from home.

A CHOICE OF JOBS

With the end of the war in Europe in 1945, and the limited prospects for WAECs my thoughts became increasingly focussed on what the next move might be. I had become more and more interested in advisory work as the war progressed, and also in practical farming as a result of running the farming operations on the Committee's land, and the feeling of achievement when things went well. I thought that I would probably look for a job in those two areas, rather than go back to education. From 1944 onwards, various odd propositions had been put to me which I believe were the result of my name having been put on a list reputedly kept at the Ministry, of people with reasonably good records of performance. I remember being approached about some strange project in China, and a much more attractive one in Jamaica, but I turned them both down as I did not want to go abroad again in the light of the Sudan experience. Then, quite suddenly, in the winter of 1945, no fewer than four opportunities presented themselves, almost simultaneously. The Government had recently decided that in view of the importance of the agricultural industry, the moment had come to set up a unified National Agricultural Advisory Service under the Ministry of Agriculture. This would replace the piecemeal and haphazard system which had operated under the universities and county councils before the war. It would provide an advisory service in every county, staffed by technically qualified agriculturists. Regional centres would be set up to provide specialist advice for the county staffs, and to carry out investigational research into local problems. Recruitment for the service had started, so I thought that I would apply, though I was not too sure that I wanted what might become a civil service type job.

Then, out of the blue, came a letter from the Ministry to say that they were shortly appointing an assistant to the Agricultural Attaché at the British Embassy in Washington, and they had put my name on a short list for interview if I was interested. This sounded quite exciting, though I was a bit dubious about moving out of the country in view of my father's failing health. However, I thought that it was too good a chance to miss, and I was duly summoned for interview.

Before that happened, Geoffrey Blackman suddenly offered me a job one day when we were out inspecting some of his plots. He had just been appointed to the Agricultural Chair at Oxford with a remit to set up a department of agricultural science in place of the old rural economy degree which before the war had the reputation for being a soft option for undergraduates wanting three easy years at the university. He told me that they would be acquiring a farm near Oxford, and that I might have a hand in running it. Then the very next morning came a letter from Dunstan Skilbeck, the newly appointed Principal of Wye College saying that he would like to meet me to discuss the possibility of a job with him. (I had met him before the war when he lectured at Oxford.)

So there I was, suddenly faced with four possible options for the future. It was quite a boost to the ego, though that did not make it any easier in deciding which signpost to follow. The first thing obviously was to clear the Washington interview, as this might well reduce the options to three, not four. That is what happened, for after a fairly exhaustive interview, the post was offered to someone else. I heard later from Robert Rae that I had been placed second because they were looking for someone with economics training – which I had not got. It was as well that I was not offered it, as my father died soon afterwards, and it would have been unfortunate if I had been in America at the time.

The next thing which came up almost immediately was the interview for the new Advisory Service, which went well, though no definite job was offered at the interview, as expected. I told the Board that I was considering two other proposals, and it was less likely that I would accept a job with them, even if it was offered, being unsure whether I wanted to become part of a state service. It looked then as if it was a choice between Wye and Oxford, unless both of them seemed unattractive, in which case I could fall back on the new Advisory Service.

The next step was an interview with Dunstan Skilbeck at Wye. Driving down, I decided that if the job looked reasonably attractive, I would go for it. The college had a good pre-war reputation, a very good farm, and a generally very pleasant atmosphere, and I already knew one or two people there. It looked a better bet than the uncertainties of Oxford, with its present lack of a farm, and its poor pre-war academic reputation, even though I found the prospect of working in the environment that I had known at Cambridge to be quite an incentive. There was also the important question of personalities. I knew that Skilbeck had rather a volatile personality, but that he was extremely energetic, and a man with progressive views who would probably be very successful in raising funds for projects at the College. On the other hand, Blackman was very easy to get on with, had an excellent reputation among his research workers, and seemed likely to be able to build up a good scientific department.

In discussion with Skilbeck, it became clear that he had not had time to develop his ideas fully, or to make up his mind which posts were on offer, as he had not yet interviewed several of his 'possibles'.

It was at that point in our discussion that his telephone bell rang to interrupt our conversation. Dunstan at the time was still a Group Captain in the RAF in which he had served throughout the war with a distinguished record. I am sure that he will not turn in his grave, if I recount what happened next – in fact, I expect that he would laugh heartily, as he had a good sense of humour. The chap on the other end of the line must have been some junior official in the RAF, I think, from the trend of the conversation. It seemed as if Dunstan wanted a warrant, or something of the kind, to get to a meeting, and this chap was not empowered to let him have one because of some regulation or other. The discussion got increasingly heated with Dunstan jumping up and down excitedly in his chair, getting more and more furious as the unseen caller obviously dug in. It finished with Dunstan virtually tearing strips off him in a very loud voice, until he slammed down the phone saying that he'd ring again later on, and that he would expect something to have been done about it by then – or else!

He quietened down, and apologised for the fracas, but it was too late as far as I was concerned, for in those few moments I had said to myself, 'If he can fly off the handle like this over such an apparently trivial matter, I think he could prove extremely difficult to work with. If I like Geoffrey's proposal, I'll go to Oxford.'

The following day I rang up Geoffrey to say that I was keen to have a further talk about his job, as I thought that I would like to come and work in Oxford, and then burned my other two boats by writing to Dunstan and the Advisory Service saying that I was making other plans.

By such trivial episodes is the course of one's life determined. I must thank the unknown caller for the next 35 happy years at Oxford.

It was a great wrench leaving Surrey and all the friends that I had made there, not least the Mattinson family who had been so good to me. It was also sad to be giving up the little farming empire that I had built up there, though that was clearly destined to break up very shortly in any case. Things were already changing rapidly, and it was better to get out, and miss the dismantling which was beginning to take place. Most of all, I regretted leaving the members of the farming community, and of the committees I had served with, and all those who had welcomed me into their homes so hospitably, and their wives who had fed me with many teas. There cannot possibly be another profession – and farming *is* a profession – which contains so many kind and hospitable people.

It has been a wonderful experience working in it, and I am still trying to repay some of the debt that I owe to it for the pleasure it has given me in life.

At the end of June 1946, my bags were packed, and farewells said, and after a three weeks' holiday in Cornwall, and a visit to the first summer conference of the recently formed British Grassland Society in Aberdeen, it was up to Oxford at the beginning of August to get settled in before the new term in October.

But before moving to that next phase of a working life, this may be an appropriate moment for a retrospective look at the achievements of the agricultural industry during those six and a half years of war, and their influence on the evolution, or perhaps it should be revolution, that was to take place in farming practice over the next 40 years.

THE WARTIME FARMING CONTRIBUTION

The Government was fortunate to have had the experience of the 1917–18 Food Production Campaign to draw upon when formulating plans for another war. It started planning in reasonably good time, when the Committee of Imperial Defence set up a sub-committee on Food Defence Planning in 1936. This was charged with making recommendations for action in the event of war, and it had already made considerable progress with proposals for food rationing and the setting up of local agricultural committees by the time of Munich in 1938. That crisis added impetus to the planning process so that by the time war was eventually declared in September 1939, some plans were in place for immediate implementation.

Subsidies had been provided on wheat, and, later, for oat and barley production, and for the application of lime and basic slag to grassland with the objective of getting it into better heart for ploughing in an emergency. In the spring of 1939, the Government introduced a £2 per acre ploughing grant for old grassland, though this had had no very significant effect by the time war broke out. Grants for land drainage had also been introduced. In addition to this, the Minister of Agriculture had got the Government to agree to the purchase of some tractors as it was realised that these might be needed to do work for those farmers not equipped for arable farming.

The year between Munich and the outbreak of war was a vital one since it provided a breathing space for the finalisation of the plans for food rationing, the setting up of Food Offices, printing of ration books, and some purchase of reserve stocks. Then the few months of the 'phoney war' which followed the outbreak of hostilities were also extremely valuable in giving the Government more time to put their plans into operation before the blockade became serious.

On the declaration of war, 61 War Agricultural Committees were immediately activated, with control over some 300,000 farmers and growers, and on 8 February 1940, the Ministry of Food was established, though the first rationing schemes for sugar, butter and bacon were not introduced until the following January. These were followed by those for meat in March, and for margarine in July.

The Agricultural Committees were given an initial ploughing up target of two million acres, of which one and a half million were in England and Wales, and each county was given its individual quota. They had wide powers under Defence Regulations to order farmers to carry out specific operations, though these orders were also to some extent to protect farmers against breaking clauses in tenancy agreements. Priority was given when ploughing old grassland to the growing of wheat and potatoes. It was, however, too late in the year to have much effect on the wheat acreage for the 1940 harvest, and much of the ploughed land went into oats and barley the following spring. District Committee members went round farms in the autumn and winter persuading farmers to offer grass fields and by the spring the target acreage had been achieved, even though doubts were expressed by farmers about their ability to do the job owing to the lack of equipment. The Government's reserve of tractors became extremely valuable, for if farmers hadn't got the equipment themselves or could not find neighbours to come in and do the work on contract, they could go to the Committee as a last resort.

A further target of two million acres for ploughing up was set for the 1941 harvest, and this was increased to two and a quarter million acres, as shipping losses were already starting to exceed Government estimates. This second year of the ploughing campaign realised an additional 450,000 acres of wheat, and 300,000 of potatoes, which gave an increase of 25 per cent of wheat tonnage compared with 1939. Further ploughing quotas were set for the next two years and by 1943 the acreage in wheat had reached three and a half million acres, as large as it had ever been, even in the 1870s. 1943 also turned out to be a good harvest year with yields 13 per cent above average, and this alone provided a welcome bonus of an extra 500,000 tons. This contribution from the ploughing up campaign not only allowed the country to get through the war without bread rationing, but also provided large amounts of animal feeding stuffs in the form of oats, barley, beans and peas, which maintained milk yields, and made a moderate contribution to the supplies of meat.

It was ironic that bread rationing had to be introduced after the end of hostilities in 1946, because of the world shortage of grain brought about by the disruption of the war on the Continent, and some bad harvests. The shortage of food in Europe was so acute in 1945–6, that Britain actually supplied one million tons of cereals for the relief of starvation. It was, therefore, not only this country that benefited from the food production campaign but other countries as well. Bread supplies had been helped early on by raising the extraction rate for flour to over 80 per cent, which provided additional flour, but naturally this led to a less attractive looking – and tasting – loaf. It also meant less animal feeds from milling by-products.

At the start of the war, there were no plans for the rationing of animal feeds for, as in the First World War, the Government seemed to be optimistically hoping that it would be possible to import enough to keep the dairy, pig and poultry industries adequately supplied, even if beef cattle and sheep had to go short. But it soon became clear after the fall of France in 1940 that the war was going to be a good deal longer than originally anticipated, and that Britain was likely to become increasingly besieged. So in the summer and autumn of 1940, hasty plans for the rationing of animal feeds were drawn up, and a full rationing scheme came into force in the spring of 1941.

The scheme only permitted an allocation of concentrate feeds to dairy cows, pigs, poultry and working horses, with dairy cows having first priority. This was on the advice of nutritionists, who argued that milk was a protective food, which could help to offset vitamin and other minor element deficiencies in the diet. Beef cattle and sheep had to rely on grass and forage crops, and home grown oats and barley. A shortage of grain led to rapidly rising prices adding further to the problems of beef and sheep farmers. At one stage, it is said that the Ministry was seriously considering a compulsory slaughter policy for beef cattle in order to reduce their numbers, but the plan was abandoned on the advice that it would be virtually unworkable, and would have a serious effect on the morale of farmers.

Allocations of food for pigs and poultry were much reduced, and the pig population dropped from about four million head in 1938 to only one and a half million by the end of the war. The Government assisted pigkeepers' clubs which encouraged people to keep one or two sows, fed mainly on household waste. Members were allowed to slaughter one pig a year and keep a side of bacon for their own use. A number of household waste processing plants were also set up to produce the famous 'Tottenham Pudding', which pigkeepers could buy to help feed their pigs on small units. In a parallel attempt to encourage domestic producers, the Government sponsored Dig for Victory campaigns, fostered poultry and gardeners' clubs and other 'backyard' activities, and put out advice on self-sufficiency through the media.

Farmers quickly adapted to the shortage of animal feeds, and a reduced grassland acreage. Initially, there was a drop in milk yields with the change from the high concentrate feeding which had become fashionable between the wars with cheap protein readily available, to a system relying on own home-grown foods, and better managed grassland. It was similar, in a way, to the relative ease with which, 40 years later, farmers adapted to the imposition of milk quotas in 1983–4.

The lowest point for milk production was the winter of 1941–2. By 1943, output was back to where it had been before the war, though by then there had been an increase of about 15 per cent in the number of producers, as some beef and sheep farmers switched to milk. In 1943, to make more milk available for the market, the Government introduced a national calf starter meal based on a dried skim milk and whey substitute to replace whole milk for calves. I was taught as a student that you couldn't rear a calf adequately on under 70 gallons of whole milk – and how ridiculous that sounds today! We ran a number of trials and demonstrations to prove that you could rear calves well on a starter, and though traditional farmers were doubtful at first, the technique caught on quickly. It was estimated that by the end of the war, some 50 million gallons of extra milk were available for the public as a result.

Wartime agricultural statistics show that the yields of the main cereal crops, potatoes and sugar beet did not decline, as might have been expected taking into consideration the fact that much of the old grass ploughed up was of very poor quality, being both short of phosphate and generally very acid. That the yields were not affected was probably due to the large quantities of lime which were applied under the generous subsidy (the quantity used increased tenfold compared with that used in the '30s), and the fact that farmers were doing better financially and were prepared to buy as much phosphatic fertiliser as they could get under the rationing scheme. The same was true for nitrogen, as farmers began to use more on grassland to make up for the acreage lost by ploughing. There was also better drainage on heavy land, due to the huge amount of tile and mole drainage done under generous subsidies.

Standards of husbandry also rose, with more timeliness

in cultivations, helped by increased mechanisation. Crop varieties did not change much, though a few new cereal varieties were introduced. These though were still prone to lodging under fertile conditions.

Mechanisation proceeded apace throughout the war, and farmers, with rather more money in their pockets, were anxious to adopt it. The Government's purchase of tractors in 1939 helped Ford to expand its production and they were given generous allocations of steel, the production of which had been greatly expanded to cope with the demands of the armed forces and shipbuilding. From an annual production of some 4,000 tractors in 1938, Ford increased their output to 19,000 by 1941. In addition, tractors were being imported from Canada and the USA, especially after the introduction of Lend Lease. The increase in overall numbers is graphically illustrated by the statistics: 50,000 in 1939 had increased to 95,000 in 1942, and 175,000 in 1946. It wasn't until the end of the war that Fergusons imported from the States first appeared on farms. Of course, it wasn't only tractors, for combines, balers, ploughs and other implements were imported or produced at home in increasing numbers. The 500 combines working in Britain in 1939 increased to 2,400 by 1944, and the number of balers rose from 1,000 to 3,300 over the same period.

There was such a demand for both new and second hand machinery, both from farmers and from the WAECs who were being called on for more and more contract work, that the Government introduced a quota system for new machines, and imposed price controls on second hand ones in 1941. The allocations were made by the Machinery Committees, which were given quarterly quotas according to the availability of new supplies. Farmers wanting to buy a new machine had to make an application to the Committee with a good case to justify it, and the Committees then had to judge as to who had the greatest need. The Government allocated something like 10 per cent of new machinery to the Committees for their own use. Inevitably, there were complaints about favouritism, but, by and large, the allocation system worked well, and it is difficult to think of any other way in which it could have been done.

The other main factor in production – labour – also came increasingly under the control of the Executive Committees, who set up sub-committees to deal with the work. The main source of additional labour required for increased production was the Women's Land Army. This reached a peak strength of 80,000 in 1942, at which point the Government restricted recruitment, since girls were also needed in other sectors of the war effort, and that figure was regarded as sufficient to meet the additional agricultural requirement. Many girls worked individually for farmers, but the majority were employed directly by the Committees who supplied gangs on a contract basis, or used them for their direct farming operations. At the peak of the war effort, the Committees operated nearly 1,000 hostels for Land Girls and refugee workers, and drew on prisoner of war camps as well for gang labour. In spite of the increasing degree of mechanisation, a huge amount of work still had to be done by hand, especially at hay, harvest and threshing time.

For two or three years in the middle of the war, many Committees organised farming holiday camps at harvest for students and urban workers. These provided useful war work, a break in the country, and some pocket money at the same time. Finally, older children should not be forgotten, for without them, it would have been almost impossible to have got the potato crops stored before the winter.

Another function of the labour sub-committees was to deal with all matters relating to the call-up of farm workers for military service, and it was their responsibility to judge whether a worker reaching call-up age was essential to the work of the farm where he or she was employed. They also had to decide, as the call up age was raised, whether workers then caught in the net were essential, or whether they could be spared for the forces without too much detriment to agricultural production. This was not an easy task, and it involved visiting the farms to assess that person's contribution, and their particular role in the farm's labour force. The committees were also responsible for placing and supervising trusted POWs who were allowed to live and work on farms individually, if they were considered a safe risk. Many of the German prisoners taken in the later stages of the war had experience in farming, and they generally proved to be excellent workers.

The overall effect of this massive effort, of which the WAECs were an essential component alongside the farming community, is well illustrated by the following figures:

Year	Food Imports (Millions of Tons)
1934–8 Av	22.0
1940	18.8
1941	14.6
1942	10.6
1943	11.5
1944	11.0

Food and Agriculture in Britain 1939–45, Hammond

Naturally strict rationing schemes had a very significant effect on imports, but in the middle years of 1942–44, it has been estimated that 5 million tons more food were produced annually from British farms than in 1939.

The vital link between the Government and farming were the WAECs, and without them, the results could never have been achieved. Part of their success was due to the fact that they were, in a way, part of the farming community, in that they were managed primarily by farmers with the aid of professional advisers of various kinds – technical, legal and administrative. They have been criticised as having been autocratic bodies ruled by large farmer dictators and upstart officials who ordered smaller farmers about and took away their farms and their livelihood; but that is a completely false picture.

Certainly the powers vested in them were draconian and far reaching and by the end of the war covered virtually every aspect of farming. But there were many checks and balances against the misuse of those powers, and they were no greater than those which were vested in many other wartime executive bodies. It should never be forgotten that it was an all out war that was being fought, and that Britain was a beleagured island, and under such circumstances some personal freedoms and rights have to be sacrificed for the common national good. As I have already said, in my experience, dispossession was only resorted to when all else had failed.

Far from being little Hitlers, many District Committee members were more akin to Florence Nightingales, generally nursing their patients with frequent visits, helping them to draw up plans of work and with the purchase of seeds and other requisites, advising on cultivations, and so on. I knew of one District Committee member in a neighbouring county who would visit a really incompetent C farmer in his area at least three times a week early in the day simply to plan his work and get his labour organised for him. That particular farmer in a different county could well have been dispossessed, but he was kept afloat by the efforts of his local Committee member. By and large, the members of the local Committees did a very difficult, and often distasteful job extremely well, and the success of the system depended on their tactful dealings with their fellow farmers, and the immense amount of time which they devoted to helping them in many different ways; and it must be remembered that they had their own farms to run at the same time with their own pressures to face. Much the same could be said of the District Officers, but they were paid for it whereas the farmers were not rewarded in any way. I have always felt that they have not received the recognition that they deserved for the work they did for their weaker brethren, and for which they often received more brickbats than thanks.

These War Years, as I have recounted in passing, encompassed a period in which many technical advances which were to transform the face of farming over the next 40 years were introduced for the first time: AI, systemic herbicides, milking parlours incorporated into farm buildings, and, near the end, the application of hydraulics to farm mechanisation through the Ferguson system – to name just a few of the major ones. Many of the new techniques could not at the time be fully exploited due to shortage of materials, or to deficiencies in supporting techniques not fully developed. The spread of AI, for example, was limited until more efficient methods of dilution and longer term storage were developed, and the use of herbicides was restricted until more efficient sprayers and nozzles were perfected. The introduction of better combine harvesters capable of dealing with heavier and moister British crops depended on the production of more powerful engines and on self propulsion.

But it was always the case that as soon as a genuinely new and promising technique was introduced, there was no shortage of progressively minded farmers anxious to experiment and develop it still further. Nor, for that matter, was there any lack of manufacturers ready to design and fabricate new equipment and machinery to support it in the field. The milking parlour is a classic example of this. In the years that followed the introduction of the three abreast double standing parlour at ground level in permanent buildings, there was an absolute spate of new designs, systems and equipment. These were experimented with, modified, and then frequently abandoned as new ideas appeared, which seemed to offer more efficient milking at lower cost. Even the 20:20 computerised models of today, which we could not have envisaged in our wildest dreams 40 years ago, are not the end of the road by any means.

The final aspect of the wartime scene which deserves to be recognised and recorded is the development in information technology. Not, of course, that there was much technology about it at that time. Ideas were mainly spread through direct dissemination of knowledge and experience, by any means possible, either through the mass media or through local information channels, often sponsored by the Committees. The Government quite legitimately used the mass media for its own propaganda, and to put across information to farmers and growers so long as it was not likely to be of value to the enemy. The BBC had an extremely active agricultural department under the able control of John Green, and the main dailies had their own agricultural correspondents giving liberal coverage to agricultural affairs. Then there were the two main farming journals: the *Farmer and Stockbreeder* and the *Farmers Weekly*, other farming monthlies like the *Dairy Farmer* and the *Journal of the Ministry*, and, of course, the local press. On top of all this, there were the Technical Development Committees organising farm walks, evening meetings, farming clubs and issuing their own publications when the occasion demanded it. It is surprising that farmers were not punch drunk at the end of it all, but they did not seem to be, for they would come to demonstrations of new machines or techniques with great enthusiasm.

It is all the more disappointing for those of us who had a hand in it then, to see the continual misrepresentation of farming issues today by what is probably a quite small, but apparently influential minority of people engaged in the mass media, who seem to have their knife well and truly into the agricultural industry.

CHAPTER THREE

Oxford – The Post-war Years

———∞∞———

THE first few weeks at Oxford were a difficult adjustment from a life full of activity, with a booked up diary, to one where there was really nothing to do but sit in a library and prepare lecture notes for an as-yet unknown audience. No steps had yet been taken to find a university farm, although Geoffrey Blackman was keen to acquire and develop one. This meant that there was no opportunity for practical agricultural activities. I knew virtually no one in Oxford, and the surviving members of staff of the former School of Rural Economy, now re-christened the Department of Agriculture, were still on holiday, so I had to find my own way around. I was much helped by Margaret Talbot, who had been the secretary/administrator of the previous school for many years. She was a marvellous character, who was extremely outspoken and stood no nonsense from anyone, from professors downwards, and she knew the workings of the university like the back of her hand. She was a good friend in those early days.

I had moved into some digs in the Woodstock Road, where there was fortunately another lodger, who was working at the Institute of Ornithology. We got on well and, after a few weeks, he heard of a small flat that was becoming vacant, and suggested that we should move in together, as neither of us were wedded to a life in digs. The flat consisted of the top floor of an old house in Ship Street, right in the middle of the city, and only some five minutes walk from the Department, which suited me fine. John Gibb had served as a gunner in Malta for much of the war, and in his spare time from manning gun sites and shooting at Italian planes, he had made a study of the birds of Malta, which he had written up in the form of a thesis. This had got him the job at the Institute. He was a wonderfully easygoing character to share with, and we spent four very congenial years in the flat till he left to get married.

There were several members of staff left over from the pre-war School who were not very much in sympathy with the new look scientific emphasis of the proposed course, but, fortunately, from all points of view, they were all due to retire within a few years. Geoffrey Blackman gradually replaced them with younger scientists more in sympathy with his ideas. My original remit was to lecture on crop production, alongside a plant physiologist, Scott Russell, who had also just joined the staff. No one had been appointed to lecture on animal production, and the only animal lecturer was a nutritionist from the previous regime.

I was very anxious to break with the then traditional university system of dividing the teaching of crop and animal production into separate years – dealing with crops in the first year, and animals in the second year. This had always seemed quite illogical as the two componenets of production are so closely interlinked. So I suggested to Geoffrey Blackman that I should start some lectures on animals at the beginning of the second term, and run them alongside the crop lectures, and he was quite happy about it. In fact, I think he was a bit relieved as it saved him the problem, in the short term, of finding the money for another lecturer, until one of the old guard had retired.

In these days of specialisation, it may seem strange that one person should attempt to lecture on both crops and stock. However, at Oxford there were both plant and animal scientists who could cover the more scientific aspects of both subjects, and it was already clear that my role was to be that of teaching the practical aspects of agriculture to provide the background environment in which the scientific principles of production could be taught. Both Geoffrey and I were in complete agreement that any graduate on completion of the course should at least be fully conversant with practical farming issues and techniques at the farmer's level – even if they had not got practical farming skills, as such. All too often in the past there had been the criticism that university graduates in agriculture might be bright whizz kids, but if they were put on a farm, they would not know one end of a cow from another. This could not inspire any confidence in a farmer from whom they had to get information for advisory, research or economic purposes.

Most universities and colleges partially got over this problem of background knowledge by insisting on a year's practical experience before starting the course. In theory, this was a good idea, though in practice many students didn't really learn a great deal unless they were lucky with the farmer for whom they worked. Unfortunately they were often regarded as cheap labour and got little help from their employers unless they were fortunate enough to live in with the family. Neither Oxford nor Cambridge had ever adopted that policy, largely because of the peculiar admissions system in both cases, where the colleges admit their undergraduates and not the university itself.

Apart from reading up recent literature and preparing lecture notes, there was one other major job to be done before term started, and that was to organise some farm classes, which would have to be arranged with co-operating farmers until such time as we acquired our own farm. The County Officer for the National Agricultural Advisory Service (NAAS), which had by then been established, was most helpful in giving me names of progressive farmers representing different types of farming in the locality. I visited these farmers so that I could assess both their farming systems, and their likely reaction to having visits from students. I found them all, to a man, to be most welcoming and co-operative, and I was extremely grateful to them for being so helpful and free with their time, and for being so good with our students when term finally got under way.

Quite apart from providing opportunities for students to visit most of the leading farmers in the district and to hear their views on farming and agricultural development at first hand, the trips also gave me personally an entrée into

Oxfordshire farming circles which was quite invaluable when I was getting established in the new job. The wartime farming clubs were still very active, and I joined one of these, and through taking part in club events quite rapidly came to make many friends in the farming community. Even after we finally got our own university farm four years later, I continued to use many of these farms for outside classes. This was both because of the facilities that they were able to offer to see things which we could not see at home, but also because of the great benefit that students can derive from discussing agricultural affairs and practice with farmers face to face.

At the end of September 1946, about two weeks before term was to start, my father was taken ill in Cornwall, and he died a week later, which meant that I had to be away settling up his affairs just as the undergraduates were arriving, and needing to get advice on the course and to fix up times for tutorials. I did just manage to get back before term began, to get some tutorial teaching organised, though it was all a bit of a scramble. I was grateful to the powers who organise these things that my father lived long enough both to see my brother again after his wartime incarceration in Changi, and also to see me getting established at Oxford. I like to think that he was pleased that his initial guidance had borne good fruit.

So I got back from the funeral to find a strange assortment of people waiting on the doorstep, some 32 of them, and all extremely anxious to get down to work and to make up for the wasted years of war – wasted that is in educational terms, though not wasted in terms of maturity and experience. Other writers have paid tribute to the exceptional quality of the post-war intakes, both after the First and the Second World Wars, and I can certainly endorse that view, for the undergraduates that came up in 1946 and 1947 were quite exceptional. The new course was to be of three years with a first year of preliminary science shared with the botany and zoology schools. The ex-servicemen were exempted from the first year so we had them into the Department straightaway for a two year course.

They were a very mixed bunch, ranging from a few

members of the aristocracy who might have been more at home on the pre-war course, to farmers' sons who intended to go into practical farming, to others who had had a taste of outdoor life during the war and had no intention of going back to office work. Finally, there were some who were attracted to the science content of the course, and who intended to go into research or advisory jobs. They had an age span from about 22 to 33, and included all ranks from ex-Lieutenant Colonels downwards, with a number of former bomber pilots. But they all had many attributes in common – maturity and experience of the world outside, a determination to learn as much as possible in their two years, and a desire for a more settled future life. Accordingly they were a delight to teach, and easy to get on with. At 33, I was not much older than many of them, and several have remained close friends to this day.

Even before the war there had been an easygoing relationship between the staff and the undergraduates in the old School of Rural Economy. This had been encouraged by Dunstan Skilbeck, one of the lecturers (of whom we have already heard), and by the lecturer in agricultural history, a larger than life character, Douglas Amery, who was still on the staff when I arrived. He was known universally as 'The General', because of his military bearing and speech, which he had obviously carried with him since his days as a subaltern in World War One, where he finished up with the rank of Major. One rather had the impression that somehow he had never quite grown up after his years as a junior officer, and service in the Home Guard during the war had helped to perpetuate his military attitude. He was not exactly a *Dad's Army* figure but not so far away from it. He had helped to found the students' agricultural club, the Plough Club, in 1919, and remained its treasurer until his sudden death just prior to his retirement in the early '50s. In that capacity he appeared not only to condone, but even to encourage heavy drinking and on occasion mindless riotous behaviour such as was common in certain types of undergraduate club between the wars.

I well remember that, when I took over as treasurer of the

club on his death, I found in the papers a bill for damage caused to the old Clarendon Hotel at a club dinner in the '30s. This included over 50 wineglasses, other glasses and china, damage to chairs and curtains, a chandelier and lavatory seats, and some more besides – and that dinner was probably not exceptional either. (I also found, incidentally, a menu for an eight course dinner at the Randolph Hotel priced at 7/6 per head, which seems quite unbelievable today – but then Lord Forte did not own it in those days!)

But for all that, The General was a very kind, pleasant, easy going man, absolutely devoted to his job of teaching undergraduates history and basic economics, and to his Freemasonry. He was very much liked as he was quite guileless and sincere – one of those Oxford 'characters', who just do not exist any longer in the modern world.

Both Scott Russell and I, as 'new boys', were keen to continue the tradition of informality, which is assisted by the personal nature of the Oxford tutorial system, though not on the same lines as The General. In a subject like agriculture we also had the advantage of having students together on farm visits and field classes, where not only do they get the chance to know their lecturers well, but also to get to know each other much better. This is important at Oxford where undergraduates live in different colleges and only come together for classes and practicals, which is a different situation to that found in single purpose colleges.

I think that one of the most noticeable and significant changes that took place over the War Years was in the form of address between individuals – not just between students and staff, but in public life generally. At Cambridge, I was always 'Soper' to my lecturers, even to Jem Sanders and Frank Garner on the Diploma course, and I was that to my first colleagues at Reading. But by the time we started at Oxford, it was always Jack or Chris when talking to undergraduates, and it wasn't long before I was Mike to them as well. Some old-fashioned people thought that this was terrible: a lowering of the status of a lecturer in the eyes of his students. They felt that, for the sake of discipline if nothing else, there should always be a suitable gap left between them.

I take the opposite view, in that I have found that it builds up a greater feeling of mutual respect, and if that is established, the question of discipline just does not enter into it. One could always exercise authority if it was needed, and it seldom was. When students are treated as mature grown-up equals, they will behave as such.

Another development, which did not come for some years, but which greatly helped to build up a corporate feeling in the Department, was the creation of a common room for morning coffee and tea, for all members from the Professor to the cleaners. Apart from anything else, it provided a good opportunity for discussing business and arranging meetings. By the late '60s, we had the reputation of being the best department in the university for integration and staff student relationships.

That first term went by very fast and I no longer had any regrets at moving back into education and away from advisory work in the field. Life was really very busy with some three hours or more a day on tutorials, preparing two sets of lecture notes, and a farm class twice a week.

The tutorial system whereby every undergraduate must spend one, or two, hourly periods each week, either singly or in pairs, with his own tutor or another lecturer is fully developed at Oxford. The usual format is for the student to be set an essay, for which references are supplied. This will be thoroughly discussed at the next session, and the victim has to supply chapter and verse for the statements made in the essay and be able to discuss the topic in depth. The system, expensive as it is to maintain, has great advantages. It ensures that students learn how to locate, and then sift and accumulate evidence from books and scientific papers, and it trains them to record the essential facts in a systematic, concise and readable form. In addition, it prepares them for thinking quickly and responding to questioning with confidence and authority, when they have to deal with questions thrown at them by the tutor. These two accomplishments can be of inestimable value later on when they move into the outside world. Because of the numbers involved in the immediate post-war years, I had to tutor in pairs, but that was never completely satisfactory, and it was better for both parties when, as numbers dropped, we could cope with the students individually.

Just before the start of the next term in mid January, it snowed hard and turned very cold, and it was the beginning of a miserable eight weeks, during which the snow never really disappeared and there was continuous frost. I had to cancel most of the farm classes, though we did visit a few animal units. The main problem was to keep warm, as there was a severe shortage of fuel and the central heating was off for most of the time. I remember sitting huddled up in two sweaters and my old Home Guard overcoat for two or three hours on end tutoring in my freezing room. It was not so bad for the victims as they were only there for an hour, and could then get some exercise, but if I had three hours running, it was sheer misery. It was a great relief when the weather finally broke, early in March, and it was possible to get moving again.

Problems With Beans

By the end of the second term, I realised that I must do something about a research project, since that had been one of the conditions of employment. I had appreciated long ago that I was not really cut out to be a laboratory scientist either by temperament or aptitude, being far too impatient to tolerate the laborious and often repetitive observations that such work involves. My inclination and skill, if I had any, lay far more in the field with growing crops rather than with test tubes. From my early days at Cambridge, I had become interested in field beans (and not just because of their wonderful scent at flowering time, though that may have helped). I had grown them from time to time on some of our sites in Surrey during the war. One of the challenging things about the crop was its inconsistency from year to year, and no one seemed to have any very valid theories why this might be.

Even if one eliminated the obvious disasters, like chocolate spot fungus and black fly aphids, there could be a high yield one year and a disappointing one the next, using the same techniques for growing the crop. It was also clear from the literature that farmers did use a number of different methods of planting and so on, and that there were quite a number of theories and folk lore, but little solid fact.

So it seemed that a survey was needed to see just how farmers did grow the crop in the Oxford area. I found that it was grown mainly on the heavier soils, as might have been expected, as winter beans used to be regarded as an integral part of the rotation on really heavy clays.

It was not a good year to start a survey since quite a number of crops had been killed out by the exceptionally hard winter, and others seriously affected, and by the time that I got to some of the fields on my list, I found them already ploughed up again and sown to barley. Only some 12 crops had survived, and I managed to visit these three times during the late spring and early summer, taking plant counts each time to try to arrive at a figure for mortality. I had previously got sowing details from the farmers concerned and their views on growing the crop. Even taking into account the poor conditions that year, it was clear that two very important factors had been sowing date and seed rate. If the crop had gone in after mid-October, or the farmer had used a low seed rate, the result was always unsatisfactory.

I continued with the survey for another two years which was long enough to prove that some of the folk lore was well founded, such as that in wet, good growing years, pod setting was poor. In hot, dry years on the other hand, even if the crop was a bit thin, pod setting would usually be good, and the yield higher. Whether pod setting was better because the plants were smaller with less leaf, or because there was more sun, and possibly the bees were more active to facilitate fertilisation, I could not tell from the survey. Another point that came across very strongly was the importance of a regularly spaced plant population. It didn't seem to matter whether the seed was sown in 7″, 14″ or 21″ rows, or even broadcast on the surface and ploughed in. It was the number of plants per given area that was most important, and there did seem to be an optimum density of plant population which could vary with the soil and climatic conditions. In the second and third years, with the help and enthusiasm of a local farmer, John Gee of Cumnor, I was able to put down a series of experiments to try to throw more light on my theories through getting yield data on populations under more controlled conditions. Two of my students, Geoff Hodgson and Peter de Jongh, who were married and living in Oxford in vacation, were invaluable assistants in collecting data and helping with the harvesting of the plots. Geoff went on to obtain his doctorate for a study of the crop, and, later still, came back to lecture on crop production in the Department.

We had a little peg drum thresher for the plots, powered by a small two-stroke engine. Plot harvesting was a slow and frustrating business, as the drum kept on bunging up with straw at the back. One late afternoon in September, when it had been particularly obstructive, I got too close in pulling the straw away, and began to thresh the fingers of my left hand instead of the crop. The first sensation was one of intense heat, not pain, and it wasn't till I looked down and saw my middle finger sticking out sideways with apparently no top on it that I realised just what had happened. It was lucky that the drum was fairly well encased, or my hand might have been dragged in. The tops of the other fingers were very badly lacerated, but, when things finally healed up, I was left with a rigid middle finger which got in the way of everything. When I went back to the surgeon for final clearance I tried to persuade him to take it off at the first joint, but he resolutely refused until I said, 'But, look here, I can't hold a golf club properly with it like this.' Fortunately he played the game, and said, 'Oh well, in that case, we had better take it off, hadn't we.' This he did, and I have not really missed it since, except for one occasion when I was attempting to demonstrate to students how to castrate young pigs, and found that I was quite unable to apply the pressure at the appropriate place!

I carried on with the work on beans for several years, publishing papers in the scientific journals, and tried to develop some new lines of spring beans, which I had concluded were a safer crop than winter sown ones. But beans are notoriously difficult to breed because of a high degree of cross pollination, and I decided that this must be left to the experts, who, over the years since then have developed some more reliable varieties.

This work on beans, incidentally, provided my first contact with the BBC, which was to develop further in the next 20 years. John Green was still in charge of agricultural broadcasting in 1948, and, having met him during the war on some publicity events in Surrey, I sent him in a script relating to the bean survey. His farming programme, if I remember correctly, went out at 6.30 on Thursday evenings and it was, of course, live – just think of agriculture being given such a regular prime-time evening slot these days! Most broadcasts then were talks, broken up with a little discussion, and the now fashionable magazine type programmes did not come in for another few years. John was a very good producer, and being a farmer himself, was very knowledgeable, so he broke into the script at intervals, asking questions, in order to break the monotony of one voice, which is boring to listeners after quite a short time. It was slightly nerve-racking to begin with, being a live programme, but one quickly forgot that once started.

I must confess to feeling a bit sad that in spite of all the hours that I spent on beans, I don't think that I left any mark on the crop; but the same could probably be said of a number of other workers who tried their hand at solving some of its notoriously difficult problems.

Foreign Tours

Reverting back to undergraduate affairs, we had all quite quickly built up a good working relationship with our students, and also with the staff of the Agricultural Economics Research Institute, which was housed next door to us, and whose staff were responsible for some of the economics teaching in the Department. The Institute had established an excellent reputation before the war, under the leadership of C. S. Orwin, and was still widely regarded for its work under his successor Professor A. W. Ashby. During the Easter vacation, one of the junior members of his staff, John Higgs (later Sir John Higgs, secretary to the Duchy of Cornwall), whom I had got to know well, said that he was planning a trip to Denmark with some of his postgraduate students in the summer. He asked if I would like to join him with some of my students to make up a better sized party. He would be responsible for the organisation. I was very taken with the idea, for Denmark before the war was held up as a model agricultural country with its small farm structure and highly developed agricultural co-operative system, which dated back to the beginning of the century. The importance of this was clear enough from the deep infiltration of Danish bacon and butter into the British market in the '20s and '30s.

I managed to collect up 15 of our students, and we set off with a party of about two dozen at the end of June, via the Hook of Holland and by train to Copenhagen. One indelible memory of our journey remains today. We had drawn up at a platform early in the morning at I think it must have been Hamburg, and a local train pulled in on the other side and disgorged large numbers of workers. To a man, or woman, they were pale and gaunt, and dressed in the drabbest of clothes. Though this was nearly two years after the end of the war in Europe, one appreciated perhaps for the first time what the ordinary average German citizen must have gone through in the later stages of the war, and one almost felt sorry for them. For what could they as individuals have done about the concentration camps, the bombing of Warsaw, and all the other evils perpetrated by the Nazis? If they had protested, they would probably have been shot or sent to a concentration camp themselves. That feeling of sympathy diminished rapidly when we were told of the atrocities committed in Denmark under the occupation. One could feel the intense hatred of the freedom-loving Danes not just for the soldiers that had occupied their country, but for the German people as a whole.

The first week of the two week tour was spent staying at the agricultural college just north of Copenhagen at Lyngby, with day trips out from there by coach. The abiding memory was of hard wooden slatted beds, covered with a thin straw paleas, and of having dry rolled oats and cheese for breakfast! The Principal, Neils Dyrbye, a confirmed Anglophile, made us most welcome, and when I stayed at the college again two years later the slatted bed had thankfully been replaced.

The main impression of the trip was one of a constant stream of visits to co-operative milk factories. I think that John had said that he wanted his party to hear as much about co-operation as possible, which interested them especially as economists; but it came a bit hard on some of my students who wanted to see as many farms as possible. By the middle of the second week, when we were in Jutland, John and I were facing a near riot as we drew up at the sixth milk factory. Two of his students simply refused to get out of the coach, and one of mine, an ex Lieutenant Colonel, said that he wasn't b★★★ well going to see another milk factory either. At that point, I decided that it was time to instil a bit of discipline into the party. 'Look, Sam,' I said, 'I am just as bored as you are at the thought of still another factory, but remember that you are a guest in this country, and that our hosts have gone to a lot of trouble to entertain us and fix up a programme for us, and they are not to know that you have seen it all already; so you are b★★★ well going to get out of this coach and show some interest, even if you don't feel it.' Military training prevailed, and he meekly got out of the coach, and I had no more trouble, even when we reached our seventh factory the day before we left!

This was the first of four Scandinavian tours that I did with our students over the next five years. We went on one more to Denmark and two to Sweden, of which perhaps the first one to Sweden in 1948 was the real highlight, and the most memorable. This was due to the quality of the programme provided for us by the Swedish Farmers Federation, and also because of the exceptional group of students, who were a real vintage lot. Swedish agriculture at the time was very progressive, especially in the southern province of Scania — much more so than in Denmark. This was probably because they had not been occupied during the war. There was a very large acreage of oilseed rape being grown in the

South, and the combination of farming and forestry in the central area was something quite new to us. In their climate, where farming more or less closes down during the winter apart from milking the cows and feeding the livestock, the woodland area found on most of the farms provides profitable employment for the farmer, and any workers that he might have, in addition to his family. In these times of overproduction of farm products and costly set-aside, it is incomprehensible to me why our Government, bearing in mind an import bill of some six billion pounds for timber products and a suitable climate for growing trees, does not launch an all-out comprehensive campaign for agro-forestry in Britain. It is true that our farmers have no past experience of forestry, but that is no reason why they should not be able to learn from suitably designed courses, or from the advisory services, as is, in fact, happening with those now growing coppice trees for biomass production.

We were very fortunate on that first visit in having a leading member of the Swedish Farmers Federation, Dag Ljungman, as guide and companion for most of the two week visit. We spent a night sleeping on straw in one of the barns on his farm in central Sweden, shaving and washing in his lake, and breakfasting in his garden. We found the Swedes to be extremely hospitable and welcoming, and they put on a number of special lunches for us, with formal speeches and so on. The colleges where we stayed arranged entertainments for us as well. The reputation of Britain was, of course, very high at the time as a result of the respect gained for its stand against fascism, and in the case of Denmark for helping in its liberation. I also appreciated then what a prestigious reputation Oxford had in European countries, and our reception may also have been enhanced by having two members of the aristocracy in the party. One of these, now a pillar of the House of Lords, nearly got us into trouble by driving his Landrover over the top of a large mound outside Uppsala, which turned out to be a revered ancient Swedish monument of some kind!

At the end of the first Swedish visit, I went on into Norway to spend a few days with an elderly shipowner/farmer, who had a beautiful estate to the south of Oslo bordering the fjord. He had landed up in Oxford on a short course to learn English and had contacted me as he wanted to take the opportunity of seeing something of English farming while he was there. I arranged a few farm visits for him, and he joined our twice weekly farm classes, where he was very popular and made welcome by our students. He farmed about 350 acres, and claimed to be one of the largest mechanised farmers in Norway on the basis of owning a combine harvester and some five tractors!

He must have been wealthier and more influential than I had guessed from his very quiet and unassuming demeanour, for he had a large and beautiful house and estate. This had been seized by the Germans during the occupation as a residence for Terboven, the Commander in Chief in Norway, who had an evil reputation. The Germans had left their scars on the house in the shape of bullet holes through some of the portraits of our friend's ancestors on the walls of the dining hall. He had repaired most of the damage, but was leaving the holes as a permanent record of German barbarism. I don't think that we, in Britain, have ever quite appreciated the hatred and resentment that was engendered in the occupied countries of Europe – or, for that matter, the hatred for some of their own people who had been suspected of collaboration.

We had a difficult example of this on our second visit to Denmark the following year. I had become friendly with a Dane who had worked in Oxford for a short time, and he accompanied us on some of our visits over there. He had been a prominent member of the underground Resistance, and when he heard the name of a landowner whom a local college had arranged for us to visit, he flatly refused to come with us to the farm. This was on the grounds that the man we intended to visit had been a collaborator during the war, and had been ostracised by the local population accordingly. This was four years after the war and it was clear that the scars would take a long time to heal.

The Norwegian farm was well laid out on very good land, with a relatively high yielding dairy herd, and equipped with the conventional huge Scandinavian general purpose barn which housed all the livestock in the winter, plus their forage, and sometimes included a residence as well. One could see the advantages of these for having everything under one roof in the hard winters, and for the conservation of heat in a very cold climate, but they did seem to be a terrible fire risk. We were told that, in fact, they did burn down occasionally with obviously devastating results. My host told me that the building costs were becoming very high as well, and he thought that their farmers would have to resort to simpler buildings in the future, even if they were less convenient in the winter months. I stayed with him again two years later after another Swedish trip, but he was far from well on that occasion, and sadly he died from cancer the following year, so there were to be no more Norwegian visits, memorable for swims in the clear, warm waters of the fjord – warm as a result of the Gulf Stream coming close to the coast.

I have already mentioned the students' Plough Club in passing, and in 1949, the then secretary, Roger Pierce, thought that it ought to live up to its name by issuing a challenge to the School in Cambridge for an inter-university ploughing match. It was intended to hold this on a farm owned by one of our students near Bedford. But this fell through, so I undertook to try to arrange for it to be held in Oxford instead. This I was able to do through the kindness of a local farmer who provided the land, and the local Ferguson dealer (it had not become Massey Ferguson at that time) who organised tractors for us. Harry Ferguson himself, who still played an active part in the company, became inter-ested, and presented a massive and magnificent silver cup to be competed for annually. Oxford won the first match, and it received attention in *The Times* in the fourth leader which posed the question as to whether ploughing should qualify for the award of a half blue! About two years later, representations were made from Reading, Wye College and Nottingham that they might be allowed to join in, and from then on it became a five universities match, until Cambridge had to drop out on the demise of their School in the '60s. I am glad to say that Oxford managed to win it on several occasions in the years that followed, in spite of its reputation as a very scientific establishment.

Those first four years at Oxford passed very quickly with a heavy teaching load, farm classes, the work with field beans, the foreign visits, and participation in the conferences of the three professional societies of which I had become a member. Both the British Grassland Society and the British Society of Animal Production had been formed between 1945 and 1947. I joined both of them immediately as I found their summer and winter meetings provided excellent opportunities for keeping abreast of current developments in their respective fields. They were also a good place for getting to know research workers and fellow teachers in other universities and colleges. I had retained membership of the Agricultural Education Association which I had joined at Reading in 1936. One of the most valuable assets of membership was the opportunity provided by the summer conferences of visiting almost every corner of the UK, and seeing farming conditions there at first hand, which was a great help in teaching. It was to be a great privilege, which I could never have foreseen at that time, to be elected President of two of these societies in later years. The growth in the conference business was one of the features of the post-war years not confined only to agriculture, and I was later to become intimately involved in it.

POST-WAR DEVELOPMENTS IN AGRICULTURE

While I had been getting established in Oxford, and not directly involved in practical farming, a great deal had been happening to agriculture in the outside world. It was a very important period for the industry in the transition from an extremely tightly controlled and regulated wartime regime to a much less rigid system where freedom of choice in matters of policy became permissible once more. There was an acute world shortage of food resulting from the hostilities and disruption of communications, so agriculture could not be left entirely to its own devices. Rationing of many commodities was still in force, so the policy of production at all costs had to be maintained, in order to get as many items off the ration as quickly as possible; both for political and for economic reasons.

One of the first priorities of the Government was to increase the supply of meat, though it was to take another six years to get it off ration completely. The quickest way to do this was through the pig industry, due to the high prolificacy of the sow, and the growth rate capacity of the young pig, which makes it possible to market pork in well under a year from conception to slaughter. It is true that pigs need an adequate supply of carbohydrate and protein feed, which were also both in short supply for a time, but imports of maize and soya bean meal picked up fairly quickly as more shipping became available with the end of the war in the Far East.

So the Government concentrated its efforts on rebuilding the pig industry which had suffered so severely during the war, with numbers reduced by two thirds, at the expense of beef and sheep, which, in the case of beef, has a three year production cycle. It also maintained its support for dairying, and for cereals, sugar beet and potatoes, though the acreage of the latter dropped very quickly from its artificially high wartime level, largely because of the high labour input.

The major event of the period was, of course, the passing of the 1947 Agriculture Act. Its basic origins were a recognition that twice in the space of some 20 years Britain, an island nation with a very high population, had become involved in major wars at a time when its agricultural industry was operating at a very low level of production. The reason in both cases had been previous Government policy directed primarily at the export of manufactured goods. This had necessitated keeping the price of food – one of the major factors influencing costs of production – as low as possible. For some 60 years it had been aided in this policy by the availability of ample supplies of wheat,

Mike Soper on a student Scandinavian tour in 1950, with his faithful Wolseley whose steering was erratic, and cooling system non-existent

The author at Oxford 1947

Experiments with pneumatic tyres, Reading 1937

The adapted dung cart equipped with dynamometer in the yard at Sonning

Close-up of the very Heath Robinson insertion of the dynamometer

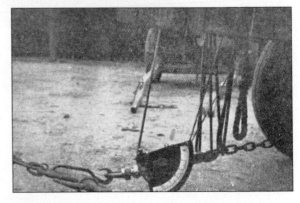

A youthful author pushing forward a poultry fold on pneumatic tyres

The canal barge converted by Cadburys for its summer school for young employees which the author helped to run in 1937

The power unit ready for his morning haul, led either by one of the apprentices or the author (seen on the right)

Three strong influences on the author's career: Professor Sir Harold Sanders

Professor Geoffrey Blackman FRS

Professor Sir James Scott Watson

Threshing sorghum (millet). The grains are beaten from the ear and then winnowed by throwing up into the wind

A bank is finished. My trusty Hamid (on the left) and workers celebrate

Undergraduates en route to Gothenburg, 1952

Swedish farmers were exceptionally hospitable. A typical scene after a farm visit.

The lighter side of Land Girl life. A sheaf tossing competition at an Open Day on our Demonstration Farm, 1945

A 'prairie buster' plough being used to plough up an old pasture for direct re-seeding. Note the wide, flat furrow slice to minimise regrowth of old grass and weeds through the seams

Tom Creyke, Chief Executive Officer Surrey WAEC, speaking to a group of farmers on re-seeding techniques on a farm walk

The team taking part in the first ever Oxford v Cambridge ploughing match at Oxford in 1952

barley, maize and oilseeds on the world market, at prices below British costs of production. If we imported these feeds from developing countries, it provided them with foreign exchange with which to buy our manufactured goods. It was a policy which from the national point of view had many advantages, but it was a disastrous one for farmers at home. It had also led, as we have seen, to the situation in 1938 when we were importing no less than two-thirds of the food that we consumed at home.

Twice, in 1917 and in 1943, the country had been brought to a perilous position by the sinking of food supply ships. By the end of the war in Europe, the Government had decided that steps must be taken to ensure that the nation's survival should never be put at such risk again. Of course, there had been the same feeling for a short time in 1918, but in the euphoric belief that the Great War really had been 'The war to end all wars', that was quickly forgotten in the urge to re-establish the country as an industrial exporting nation. Hence the repeal of the Corn Production Act in 1920, followed by the calamitous collapse in farm prices.

With the history of the past 30 years to guide them the politicians in 1945 were more realistic – both those who had served in the wartime administration, and those who succeeded them in the Labour Government, which in some cases were the same people, such as Clement Attlee and Tom Williams, the new Minister for Agriculture. He had served with distinction in partnership with Robert Hudson, the wartime Minister, and was widely liked and respected both in Government and in the farming community. Discussions with influential people in the industry had started, I believe, as early as 1944 with the aim of mapping out a policy which would guarantee some future security for the farming industry. So Tom Williams already had the blueprint for an act, when he took over as Minister in 1945. A White Paper was presented to parliament in November 1945, and, with the support of all parties, the Act was on the statute book by 1947. Its easy passage was partly a recognition by parliament and the public of the tremendous efforts which had been made by farmers and the industry as a whole to produce enough food to keep the population reasonably well fed over the previous six years. There has probably never been a time when farmers were so popular with the public. Of course, it was far too good to last, as it was not to be many years before they were being accused of being feather bedded by the payment of excessively large subsidies. But it was good while it lasted.

The Act was, in fact, very comprehensive, covering a wide range of ministerial and other activities, and tidying up a lot of loose ends resulting from the wartime powers of the Ministry. But it is today chiefly remembered for, and its place in history depends on, that section relating to guarantees to support farming. It is perhaps worth quoting these, as they were to be the cornerstone of Government policy for many years – until 1973, in fact, when Britain acceded to the then E.E.C. and had to jettison many of its domestic policies in the cause of European integration.

> The objective of this policy is to promote a stable and efficient agricultural industry capable of producing such part of the nation's food as in the national interest it is desirable to produce in the United Kingdom, and of producing it at minimum prices consistently with proper remuneration and living conditions for farmers and workers in agriculture and an adequate return on capital invested . . . The twin pillars upon which the Government's agricultural policy rests are stability and efficiency. The main method of providing stability is through guaranteed prices and assured markets. Stability is not enough. It is equally important to ensure that the industry attains the highest possible degree of efficiency.

It was proposed that the main mechanism for ensuring these objectives would be an annual price review, at which the Government would sit down with the representatives of the industry to determine the schedule and levels of prices to be guaranteed for the following year. This was the birth of the Price Reviews which played such an important role in shaping the structure and efficiency of the industry for the next 25 years. It meant that by the time we entered the Community, we did have, at least, a well equipped and vibrant farming industry with an up-to-date structure, which compared very favourably with our new competitors.

The Act also covered the conversion of War Agricultural Executive Committees into County Advisory Committees, measures to ensure the preservation of good agricultural land, independent land tribunals to hear appeals, and the creation of an Agricultural Wages Board. So it can be seen that it was an impressive piece of legislation which was the most far reaching of any previously adopted for the control of agriculture. Though many of the provisions have since been nullified for one reason or another, some still influence the industry today, such as the Wages Board.

The effect of the Act on the industry was far reaching, for it provided, for the first time, a sense of security. There was the feeling that it was safe to invest, and indeed borrow money to invest, in improvements to land, buildings and equipment. It was safe also to embark on new enterprises or adapt existing ones to exploit the technical advances that were now coming through from research, both fundamental and applied. There was a buoyant feeling among progressive farmers that at last they could go ahead with expansion without the fear of sudden switches in policy which could, overnight almost, bring ruin to those who had borrowed money to enlarge their activities.

From the Government's point of view, the effect of the Act was also beneficial. There were, as we have said, still widespread shortages of food, and, politically, it was important to achieve de-rationing as soon as possible. The confidence engendered by guaranteed prices, improvement grants (also covered in the Act), and the spur to the adoption of new practices all helped to increase production. There was still a very serious shortage of foreign exchange to buy food from abroad, so it was obviously important to produce as much as possible at home.

The annual review system also gave the Government the opportunity to apply a gradual squeeze on inefficient producers through price controls to levels at which the incompetent could not compete, but which were still adequate to allow efficient ones to make satisfactory profits. There is no doubt that over the next 25 years, the manipulation of prices at the reviews did help very materially to raise the general efficiency of the industry, and led to the gradual disappearance of the inefficient or very small producer. Of course, the unions always protested at the extent of the squeeze, and sometimes had to submit to the imposition of an unagreed settlement.

But one always felt that the larger and by now better equipped farmers were not too concerned at the gradual elimination of those who could not compete, as they themselves were still more than capable of doing so. They still had money over for reinvestment for the future, and perhaps for times when the returns might not be so good.

So the passing of the Act heralded an era of unprecedented prosperity for a large proportion of farmers, and this in turn rubbed off on the ancillary industries which provided goods and services for the practical farming sector. Research, both at Government and commercial company level, flourished, and farmers were eager to exploit the new technology presented to them from the widespread investment in research and development. The Agricultural Research Council extended the range of the research stations and expanded some of those already in existence, while the Ministry set up its chain of Experimental Husbandry Farms to test out, at the development stage, the findings emanating from basic research. Large commercial companies, especially in the pharmaceutical field, set up their own research facilities. This was an era when it was still thought inappropriate that independent research bodies should accept money from commercial companies, as people might think that any results could be in some way tainted.

Mechanisation was beginning to spread more rapidly in the practical farming sector as farmers had more money to buy machinery, and the employed labour force began to fall as a result. It was the beginning of a steady decline in manpower that has continued to this day. In a way, agriculture has been lucky in that the decline has been a continuous steady draining away of labour from the farms, rather than the sudden catastrophic annihilation that has occurred in, for example, the coal, steel or shipbuilding industries. It has made the adjustment more painless, both for those that have left the industry, and even for the employers remaining in it. It was very noticeable at the time that the loss of agricultural employees started much sooner and was more drastic than the loss of employers – the farmers – for whom the decline started later on, in the '60s and '70s. This was due to the structure of the industry with many small farmers employing little labour, who were therefore able to hang on longer, as the going got more difficult. In the immediate post-war period, it was very much a matter of adapting to technical change, rather than structural change, which was to begin to gather pace in the next decade from 1950 onwards. This was brought about partially, of course, by the adoption of the new technology by the larger and more progressive farmers, which made it more difficult for the smaller farmers, possibly farming poorer land, to compete.

A study of the statistics for the immediate post-war years brings out some interesting points. Contrary to what might have been supposed, the ten years from 1946 did not show much increase in the acreages of the main crops, nor a very significant increase in the number of dairy cows. In fact the acreage of cereals and potatoes fell for five years as the industry returned to a more balanced system between crops and livestock after the feverish ploughing campaign of the War Years. Land was put back to grass, and the numbers of sheep and beef cattle began to increase again as more grassland became available to sustain them. A most significant change was the beginning of the dramatic decline in the acreage of oats. This plummeted from a very high peak during the war, when they were needed both to feed horses and also to provide homegrown carbohydrate for dairy herds as well. Barley began to take its place as a feed crop for ruminant animals, though it would be another ten years before barley really came into its own as the major cereal crop, in the 'Barley Baron' era.

But the overall statistics of acreages and number of animals do not tell the whole story, for although acreages did not change much, yields per acre certainly began to increase, as did yields per cow. Newer varieties of cereals came on the market with stiffer straw which enabled larger amounts of fertiliser to be used. New herbicides were bringing about a revolution in weed control which allowed crop plants to utilise a higher proportion of the nutrients available in the soil. Another factor contributing to higher yields was mechanisation, not just in the range of machines available, but in their increased rate of work.

The ability to work the land speedily on difficult farms when conditions are fit, and to prepare fields for drilling by the optimum sowing date has been one of the most important factors in increasing yields since the War Years. In the days of horses, and indeed of the earlier low powered tractors, there were always fields on heavy land farms that just could not be worked down to a satisfactory seedbed quickly enough for both spring and autumn sown crops. There are still problems, of course, on clay land farms in wet seasons, but today's farmers probably don't realise how fortunate they are to be spared the frustrations of late drilling, and the knowledge that each day lost, or each crop puddled into a poor seedbed was bound to lose them money at harvest time. Of course, the cost of the equipment needed to keep up to time gives them a different sort of headache in relation to their bank manager, but I think that I would prefer that to the impotence of being unable to get on with the work when you needed to.

CHAPTER FOUR

Years of Expansion in Farming – 1950–60

———— ❦ ————

CHANGE is continuous, but, in chronicling it, reference points are needed against which events can be recorded and time scales measured, and the start of each new decade serves that purpose well.

SETTING UP THE FARM

At Michaelmas 1950, we finally gained possession of a university farm, though this was a year later than it should have been. The delay had been caused because the university land agent failed to give the sitting tenant his notice to quit by the required date, and there was nothing for it but to start again the following year. The university owned two estates, Nuneham Courtney and Wytham, in the vicinity of the city, both of which had been acquired quite recently. The former was bought primarily for its woodland; the only farm which might have been obtainable was quite unsuitable for our purposes. It had been agreed that we must have a farm on one of these estates, in order to avoid having to purchase one on the open market – even if a suitable one could be found.

The Wytham estate, lying just beyond the western section of the Oxford ring road and the Thames, had been purchased in 1942 with the object of preserving a large area to the west of the city from building and industrialisation. It was felt that Cowley to the east was quite enough to be going on with! A reasonably large sum towards its purchase had been raised by public subscription, I believe, and the rest of the money had come from a trust fund. So there was an obligation to use the estate responsibly on amenity grounds, and any development was severely restricted, though this fortunately did not apply to farm buildings.

The estate comprised seven smallish farms and some 600 acres of woodland on a ridge dividing the farm land into two quite distinct blocks. The vendor and his wife lived in the historic Wytham Abbey on the edge of the small village, and they had lost their only daughter rather tragically a few years before. It was said that they had made a fine public gesture by selling the estate to the university for much the same price at which they had bought it 20 years or so previously. This may well have been the case, but what was not mentioned was that, in the intervening years, practically nothing had been spent on the estate, either on the farms, the woodland or the village. Consequently the university inherited a vast backlog of expenditure, especially on the buildings.

The largest farm on the estate was of 310 acres, and was really the only one that appeared to be remotely suitable for what Geoffrey Blackman and I had in mind – he for his experimental unit, and I for teaching purposes. A major advantage was that it was tenanted by an elderly man, and was in a quite deplorable condition, so that there was likely to be no difficulty about getting possession on the grounds of bad husbandry. The tenant had been classified as a C grade farmer throughout the war, and had been under continuous supervision of the War Agricultural Committee. The condition of the farm when we took over was certainly as bad as anything that I had seen before, even when dealing with wartime dispossession cases.

The farm was made up of three distinct areas. There was a low-lying alluvial flood plain, with about three-quarters of a mile of frontage along the Thames. This extended back to a raised gravel terrace at a distance of some 200–300 yards. The farm buildings were situated on the gravel, as one would have expected. The terrace, very sharp and dry in places, stretched back another few hundred yards on average, and then merged into an area of very stiff, poor draining Oxford clay as the land rose up to the edge of the woods which crowned the hill. The farm was triangular in shape, the apex being at the west end, where the woods came down to meet the river. The base of the triangle adjoined two other smaller farms, which had their buildings in the village half a mile away.

The farm buildings, consisting of a number of cattle boxes, a stable with loft, a small cowshed, and a very large barn and a Dutch barn, were grouped on three sides of a square, with the farmhouse and an outbuilding completing the fourth side. They were all very dilapidated, most of them with weather-boarded walls. In the small cowshed, some of the boards had slipped or rotted and the cows were milked with the wind whistling through the gaps. It must have been extremely uncomfortable for both man and beast in the middle of a cold winter. The barn was a very large and attractive building, which stood out amidst the general signs of decay, as it had been beautifully re-tiled shortly before the war. They had obviously not repaired the weatherboarding which clad its sides at the same time. At the back of the stable was another partially roofed cattle hovel, knee deep in farmyard manure, and the whole of the middle of the central yard was full of weeds, discarded implements and piles of manure thrown out of the sheds surrounding it.

There was no piped water supply or electricity, the nearest point for each being in the village. The only water available in the buildings was that from a deep well in the farmhouse kitchen. It was pumped up by hand daily and was used for the house, for cooling the milk, and as drinking water for the livestock if this was not available from rainwater tanks round the buildings. The cows were left out all the year round, so the herd had access to the river for drinking. The two farm cottages did have wells of their own but no sanitation. The herd of 24 motley shorthorns was milked by hand in the shed and two cattle boxes, and the milk then carried in buckets back to the farmhouse for cooling. It was the only time that I saw wooden shoulder yokes like those seen in museums actually being used on a farm.

It was difficult to believe that such conditions still existed in the middle of the 20th century, and only three miles from the centre of Oxford. But I don't suppose that they were particularly unique, since the years of depression had left landlords very short of money. If they also had a tenant in whom they had no confidence, it is understandable that they might have been reluctant to spend capital on improvements until a better one came along. There used to be an old saying that bad landlords get bad tenants, and there was probably a good deal of truth in it. One cannot really blame the university for the state of affairs, as it had not owned the farm for very long. However, it does have to be said that it appeared to have done very little in the way of improvements to the other farms on the estate either. The conditions were really an indictment of the previous owner for failing to keep his property in repair, and of previous governments over the years for doing virtually nothing to improve the profitability of farming. Probably, too, the land agent who was near retirement, shied away from the gargantuan task of pulling the estate round. He did not inspire me with any confidence, after he tried to tell me that the farm wasn't really in too bad a condition at all. We had a stand up row in the middle of the yard just after we had taken over when I told him that the farm was a disgrace to the university and that he ought to be ashamed of it, which somewhat naturally he resented! I was provoked by his attitude which suggested that I knew nothing about farming. Fortunately we had no dealings with each other after that, for Geoffrey organised it that the farm would be regarded as a laboratory, which brought it under another department for the next 20 years or so.

About 140 acres was in arable, much of it ploughed out of grass under compulsory orders during the war. It was in a filthy state, the light land being a thick mass of couch grass and annual weeds, and the clay land full of creeping bent, thistles, buttercup, catmint and wild oats. The soil structure on the clay had been completely lost, leaving it like putty when wet, and concrete when dry. The previous year's crops had not been threshed when we took over at Michaelmas, and the yields finally averaged 9 cwt of so-called grain per acre, but much of that was very thin and full of wild oats. Geoffrey's research team put down some herbicide plots the following year and plant counts in the spring showed wild oat populations of one and a quarter million plants per acre. As this was in the days before specific wild oat herbicides had become available, it was clearly going to be quite a challenge getting them under control. I quickly told Geoffrey to inform the University Chest that there was no way in which one could expect any profit to be made for at least two years, and that it would take much longer than that to get things into proper shape.

The first task was to determine overall strategy, and to get building work under way as soon as possible. We decided on a dairy herd of 40 cows, rising to 50 when home bred heifers became available. We also planned to keep all young stock, the heifers for breeding and the bull calves as beef finished at two and a half to three years of age. As a supplementary animal enterprise, I opted for pigs, with an initial target of 20 sows. In addition to reconstructing the old buildings to provide a milking parlour, calving boxes and young stock accommodation, we were going to need yards for the cows (no one had yet thought about cubicles), a machinery shed and repair shop, a large area of concrete and a proper drainage system. We also wanted four new cottages, and a renovated farm house fit for a good farm manager to live in. Geoffrey needed a field laboratory, glasshouses and another cottage to house a technician to look after the laboratory. He was already the Head of the Agricultural Research Council's Unit of Experimental Agronomy, which essentially embraced research into herbicidal weed control and investigations into new crops which might have an economic future in Britain, such as oilseed rape, forage maize and oilseed poppies. The Research Council put up the money for building the laboratory, a cottage and the other research facilities, and the University Grants Committee agreed to put in some £60,000 for the farm buildings and new cottages.

We used the money to convert a couple of cattle pens to a milking parlour, and the old cowshed into calf pens and boxes for young stock. The stable became a workshop and the hovel behind it a large covered machinery shed. Then behind the buildings in the main square, we put up yards for 40 cows, two bull pens, and, on a separate area, farrowing pens for the sows and fattening pens for young pigs. We also put down a large area of concrete and adapted the inside of the barn for grain bins, and built a sack drier outside. There were, of course, continuous flow driers on the market at the time, but most of the combine harvesters were still sack models, and with the harvesting of experimental plots very much in mind, a sack drier seemed to be the best bet for our circumstances. This consisted of a raised concrete platform with 48 sack sized holes covered with grids on which the bags were laid, and hot air blown into the cavity below, which filtered through the sacks drying the grain as it went. It would take anything from 2–6 hours drying depending on the moisture content of the grain. It worked well, but, of course, the labour input was very high, and one was continuously taking people off other jobs to change the sacks.

Looking back on all this now, one is struck by the smallness of one's thinking at the time; for example, only planning for 40 cows, or 20 sows, using sacks instead of thinking of bulk handling, building bull pens, when AI was already well established and, perhaps above all, not thinking of labour saving, or using bigger machines in place of men. It really was very shortsighted, but in defence, it can be said that labour was still quite cheap, tractors and implements quite small, and many of the labour saving devices available today had not been conceived. In a way I suppose that one was still geared to traditional pre-war methods. It was not to be long before I realised that we had been too small-minded, and was regretting some of the buildings with which we had saddled ourselves.

The milking parlour was a case in point. The herringbone principle had not yet arrived from New Zealand, and the thinking was still on variations on the abreast pattern, dealing with six or eight cows at a time. I got it into my head that it must be basically wrong to be doing three things in an abreast parlour – washing udders, feeding and milking, and that the herdsman's job should essentially be milking his cows, and not messing about with washing and feeding at the same time.

The building that we had to adapt for the parlour was rather narrow for the conventional three abreast double standing layout, so I put in three washing stalls at the back leading into three milking standings, all at ground level. I also thought that the higher yielding cows would not have time to eat their concentrates if only in the milking stall for some five minutes. So I decided against feeding in the parlour at all and put in a feeding passage with self tying yokes at the back of the straw yards. The herdsman could then push his feed trolley down the back of the yards and feed each cow separately. This suited my policy of using home mixed concentrates for the herd, which I was convinced was much more economical. The plan was all very well in theory, but it was labour demanding, as it meant employing a boy (or, for a time, a girl), to stand at the back of the parlour just bringing in three cows and washing their udders. Quite apart from being a soul destroying job, it was not a full-time one, nor was the herdsman's job full-time, just milking the cows, and he had the extra work later on of feeding the concentrates in the yard twice a day. We persevered with the system for a couple of years, before dispensing with the boy and the washing stalls, which then became queuing stalls, from which the herdsman let through his cows when he was ready for them. The washing stalls became veterinary treatment and AI tie-ups, and remained so when the parlour was re-designed ten years later, so they came to serve a useful purpose after all. I learned the lesson that it often does not pay to be an innovator unless one has a very perceptive mind and a deep purse. It is safer and cheaper to let other people learn by experience, but be ready to go in when a new technique appears sufficiently well proven to be adaptable to one's own conditions.

Another lesson that I learned very quickly was the futility of relatively small farmers trying to keep their own bulls with the objective of breeding brilliant performance offspring. Of course, with my early experience of the development of AI, I should have known this already, and not wasted precious money building bull pens. In the event, I only ever bought three bulls before I saw the light, and their progeny were certainly no better than those we obtained from using 'bull of the day' AI semen. We helped our AI station to test their young bulls for many years and the bull pens became much needed calving boxes as the herd quickly expanded beyond its initial 40 cows.

Both Geoffrey Blackman and I appreciated that the role of a farm manager on a university farm, or for that matter on a college training farm, can be a difficult one. In the case of the university, there is a serious danger of a conflict of interest between running a farm profitably in economic terms for the benefit of teaching students, and, at the same time, providing often quite sizeable facilities for experimental work. A manager can have a very thankless task if he is being frequently asked to divert tractors, implements and manpower to preparing land for experiments, applying treatments or harvesting, just at a time when the weather conditions are fit for pressing ahead with the rest of the farm operations.

There is no easy solution, but a farm manager with an interest in research and development can be a great asset. We decided that as I would be playing a very active role in management it would be essential to find a youngish man still making his way up the ladder, rather than an older man too set in his ways to accept interference in day to day policy. It was also agreed that the main research would be concentrated in Geoffrey's own unit of agronomy, for which he already had a good team in place for doing much of the detailed work. His main demands would be for land, basic cultivations, and harvesting of plots. We decided not to embark on large-scale animal research since this is very financially demanding and best left to properly equipped research stations; in any case, it would have been difficult to raise money for it in addition to that needed for Geoffrey's unit. The animal units on the farm could then be treated as commercial undertakings capable, one hoped, of making sufficient profits for reinvestment to improve the buildings. Most of the arable and grassland could also be farmed commercially for teaching purposes. We also agreed that all demands for research facilities from members of the team should normally be routed through me, so as to protect the manager from constant small requests from individual research workers, which was the policy I had adopted on the Sudan research farm to protect Awam Nimr.

I knew just the man I wanted for a manager, if I could persuade him to throw in his lot with us. I had to let him know that things were going to be quite chaotic for a year or two, and that it might be a few months before we could make a very cold and draughty farmhouse habitable for him. His name was Jim Paterson, a Reading graduate, who had married Hugh Mattinson's eldest daughter some two years previously, and I had visited them on the farm he was then managing in the Midlands. He was extremely capable, with tremendous energy and progressive ideas — just the sort of man to get to grips with a run-down farm. Fortunately he decided to risk it, as he relished the challenge of pulling the farm round. This he did most effectively in the three years that he stayed with us, before moving on to a much bigger farm in the Cotswolds. We then took on another young Reading man, Ian Bare, who carried on where Jim had left off for a further three years, when he had the chance of getting a farm of his own. That was one of the disadvantages of employing youngish men — if they were good, they would move on to better things rather too quickly, and after one more experience of this, we changed our policy, and went for someone more mature. This was Ian Morton, one of our post-war batch of students with whom I had kept in touch. I thought that we could probably work in harness which turned out to be the case, for he stayed until we both retired in 1981.

I always regarded a farm manager's job as neither easy nor secure, and never advised students to embark on that career unless they were very determined and dedicated. There are, of course, many excellent management jobs, probably more good ones today than in the '50s and '60s. There were then more smaller estates, and private owners who were buying into farming for investment, country status or reasons other than a pure interest in farming as such. But these owners could be short-term birds of passage, who might sell up if City investments went wrong, or whose wives didn't like the country, or whose sons grew up and wanted to take over the farming. There were many factors of this kind, and a farm manager could suddenly find himself out of a job almost overnight. In other cases, the manager would be promised a lot on appointment, only to find that promises of rebuilding were not fulfilled, or that he was treated as little more than a farm worker.

Today, with management companies and the larger land agent firms in control of very large areas of land, there are certainly more management jobs available, and it is a more secure profession with a better ladder of promotion. Even so, there is always a risk of changes in financial policy, and the manager's position under such conditions can never be entirely safe. But that, I suppose, is also the case in the business world, where hiring and firing seems to be part of the game, and no manager can feel safe unless he delivers the goods, year by year, to his board of directors. As farming becomes more and more of a big business enterprise, managers will be subjected to the same rigorous discipline as managers in the commercial world.

There was nothing for it but to summer fallow about half the arable acreage in the first year, in order to get rid of most of the worst of the perennial weeds, especially on the clay land, and we fallowed most of the rest of it the following year. On the lighter couch infested land, we pulled out literally trailer loads of the stuff, which was carted off and burned. But neither treatment made much of an impact on the wild oats, so I decided to resort to the treatment that I had been taught at Cambridge for dealing with black grass (of which for some reason the farm appeared to be free). This consisted essentially of preparing false seed beds to encourage germination, and then ploughing them in again to destroy the germinated seedlings. We had both winter and spring germinating strains of the weed, so we began with an autumn seed bed, followed by another in the spring. Of course, it meant drilling spring barley afterwards later than one would have liked, and a lower final yield, but after a couple of years the worst fields were definitely near to being under control. I remember getting overconfident that the problem was licked on one field and putting it into winter wheat, only to find that a lot of wild oats were left. Once a soil bank of seed has been built up, there is always a risk of sudden reinfestations, possibly due to slight variations in ploughing depth, which means that fresh layers of soil and dormant seeds are exposed.

Over the years, we got the level down to manageable proportions, though it was not until the introduction of Suffix in the late '60s that we really got the wild oats beaten. Then, of course, along came sterile brome in its place, to create new problems.

Geoffrey Blackman had established very good relations with the Secretary of the University Chest (Geoffrey was a very astute university politician), who fortunately lived in Wytham village, and was therefore interested in what went on there. The Chest is essentially the Treasury which controls the university's finances, and it is obviously important to get on well with those who control the money! I was always happy to provide a load of farmyard manure to the Registrar of the university, and the Secretary of the Chest, if required, but drew the line at anyone else. After the initial infusion of funds from the Grants Committee and the Research Council, all the expenditure on research was financed out of Blackman's grants, but the farm was given its own account within the Chest, quite separate from the departmental account. This was on condition that we employed a well-qualified bookkeeper in the farm office. It was a great advantage to have everything under virtually direct control, and we found an excellent book-cum-farm-record keeper in Jack Brady, who had worked in the Chest for a time, and knew the ropes. He gave devoted service to the farm for the next 25 years, when he tragically died from a massive coronary.

The university is a charity, so profits were not taxed, but unfortunately MAFF knew this, and refused to allow payment of any farm improvement grants. We could, however, draw the normal subsidies on crops and stock to which farmers were entitled. Whether we gained or lost on this, I was never quite sure, but it probably worked out that there was not much difference in the end. Over the next 30 years, we invested very large sums of money in building improvements and land drainage on which we could have drawn very sizeable grants. These were all done out of profits made from the farm, and not at the direct expense of the university, which gained a tremendous amount in increased value of the property – though this could never be realised commercially.

We built up a fairly large overdraft on the Chest in the first two years in which we were pulling things round, but we very nearly broke even in the second year, when the dairy herd was coming up to strength, and the fallowed land began to pay dividends. The Chest kindly agreed that we need not pay interest on the overdraft, and this arrangement persisted for some 15 years, until there was a change of personalities, and financial conditions became tighter. It then began to demand its pound of flesh in the way of interest. It also imposed rental charges on new land that was taken in as our two neighbours either retired or died, and the acreage eventually increased to 620. I seldom allowed our account to get into the black at the Chest, as there were always new buildings to be provided as the farm units expanded, and old ones had to be brought up to date. So as soon as I could show a decent profit on a year's working, a request would go into the Chest for permission to spend a few more thousand on the next item of equipment that reached the top of our priority list. There was an unwritten agreement that we would pay off the cost of a major expansion within three years, and we never let the Chest down on this. In return, as long as we delivered the goods in the shape of annual profits, it was quite happy to provide the modest overdraft. It was an ideal arrangement which worked very well when, from the early '50s to the middle '70s, it was relatively easy to make reasonable profits out of farming.

Mention was made of bringing in more land as our neighbours disappeared, and the Wytham estate is a classic example of the change in the physical structure of farming over the past 40 years. In 1950, when we took over the farm, there were the seven farms on the estate, several of them quite small, and all of them with dilapidated buildings unsuited to modern farming requirements. Today, there are just the two main farm units: one, the university farm occupying over 800 acres on one side of the hill, and the other of about the same size, comprising the land on the other side. Both are fully equipped with modern buildings and equipment to allow them to be competitive under 1990s farming conditions.

TECHNICAL CHANGE IN THE '50s

In looking back at this decade, it seems in retrospect that change in the livestock sector was even more significant than it was in arable farming. This was due in no small measure to the revolution that was taking place in grassland management, and the application of mechanisation to the crop. I use the word 'crop' advisedly for this was the era in which grass really began to be regarded as a crop alongside, for example, wheat or potatoes. Earlier it had been seen as something that just happened to be there which could be left to its own devices for grazing or cutting for hay. It is true that Sir George Stapledon had been preaching the gospel of grass as a crop since before the war, but it was not until the 1950s that the concept was fully accepted, and management practices began to change from set-stocking to controlled grazing systems.

There was one area of intensive grassland activity in the '30s, namely grass drying, where production *was* closely controlled. This had been stimulated by work at the ICI Research Station at Jeallots Hill on the growth of the grass plant and its response to defoliation and competition. This had led to the realisation that grass cut young and quickly dried gave a product that was almost equivalent to a high protein concentrate for the feeding of livestock. As a result, the agricultural machinery firm of Wilders in Wallingford had been approached. They designed a machine which would cut the grass with a more or less conventional cutter bar, and then lift it on fingered slats to be discharged into a trailer alongside. Clyde Higgs, the well-known Warwickshire dairy farmer, was one of the leading pioneers who adopted it in practice. The principal method of drying was on trays through which hot air was blown from a burner. It was a fairly labour intensive system, and, of course, the younger the grass the more the water that had to be driven off, but the higher the protein content. There was a lot of interest in the system, which was also used for drying grass cut from aerodromes. After the war, continuous flow driers were developed, but the costs of driving off the water on top of the increasing costs of production, harvesting and drying gradually made the system uneconomic on the average farm.

Some industrial plants, however, have survived. They grow lucerne as an additional crop to provide a continuous supply of green material over the season. The product is sold to feeding stuff compounders for incorporation into concentrate rations for cows, pigs and poultry, where the high vitamin content of the dried grass is especially valuable.

The Wilder Cutlift was really the first machine developed for bulk grass handling and was also used by some of those who were trying to perfect silage making techniques in the '30s. The story of silage making from its introduction at the end of the 19th century is a fascinating one, which illustrates very well a concept that runs through so much of technical progress in agriculture. It is that a breakthrough in one sector often has to wait for a development to take place in quite a different field before it can be fully exploited. In the case of silage and the handling of grass in bulk, it was the need to wait for increased power from the tractor before efficient methods of cutting and conveying could be developed. The same was true, of course, for grain handling.

Up to the war, silage making had not become popular with farmers for two main reasons: its high labour requirement, and the unreliability of the product, which was frequently of poor quality, due to the methods of storage in use at the time. It was made in towers, stacks or trenches, or occasionally pits excavated in the sides of a slope. The best silage certainly came from towers, and we had a large concrete one at Reading when I worked there, which was used each year and which did produce good silage. It was made from an arable oat and vetch crop which was also what we had used at Cambridge. But the labour input was very high both in getting the crop in from the field and then feeding it into the cutter blower to lift it up into the silo. I remember forking the crop onto trailers in the field, and it was a terribly heavy job, and a very slow one into the bargain.

Silage made in stacks frequently suffered from severe wastage down the sides, and this again involved a lot of labour putting it on an elevator to stack it. There was not so much lifting with trenches or pits. Again though, wastage on top and down the sides was usually considerable, even when earthed over, which was the common method of trying to get a seal to keep the air out. It was not until polythene sheeting was developed in the '50s, that the problem of wastage in anything other than a tower was solved. That was really a turning point in silage making.

Sanders and Garner had reviewed the state of silage making in a most interesting paper at the third Oxford Conference in 1938, and concluded that the limiting factors in its spread were the high labour input and the lack of proper equipment for handling the crop – a prescient statement in the light of what was to happen later.

The first breakthrough was the Paterson Buckrake, developed in the late '40s by that outstanding farmer, thinker and innovator, Rex Paterson. He built up a very large dairying enterprise, based initially on outdoor milking bails, controlled grazing and grass silage, in units of approximately 100 cows. At one time, he had, I believe, ten of these, and he kept the most meticulous records, which he used to very good effect in the management of the system. It was ultra simple and it worked extremely well at a time when both land and labour were still relatively cheap and available. The buckrake played a vital role in simplifying silage making and we used one when we started to make silage in 1951.

The next breakthrough was the flail harvester in the mid-'50s, which employed a principle used in the Wilder straw chopper. The great advantage of the flail was that it not only cut the grass fast, but also blew it up direct into the trailer. It tended to lacerate the grass slightly which led to a quicker release of the sugars in the clamp, which speeded up the fermentation.

The advent of the flail harvester more or less coincided with the adoption of the above ground clamp as the standard method of silage making. It was aided by the introduction of portable wooden sides for clamps, which made for much safer operating conditions. This was quickly followed by the specialist silo barn, with its built-in concrete walls, and in many cases the acceptance of self feed as a satisfactory way of getting the silage into the cow at a very low cost.

Before I reached that stage on the farm in the '60s I had a brief flirtation with a new method which had been tried in New Zealand, and had been made possible by the production for industrial usage of polythene sheeting. This was vacuum silage. The principle was to extract virtually all the air from the grass which was completely enclosed within polythene sheets. With the extraction of the air, fermentation was very drastically reduced, so that there was less loss of nutrients, dissipated in useless heat and waste gases.

We were fortunate to have living in the area a representative of the firm making the sheeting, whose job was to popularise the use of polythene in farming, and we were able to get hold of very large sheets through him for experimental purposes. The procedure was to lay a large sheet on the ground, and build a clamp of grass on it. At the end of the day another sheet was put over the top, and the edges of both sheets joined with a simple but ingenious seal. A valve was taped into one corner, and all the air between the sheets extracted with a tractor operated vacuum pump. Then the valve was sealed. The whole clamp would become hard with the sheets pulled tightly over the grass. By the following morning, the sheets would be blown right out and full of gases generated during the night. These had to be let out, and it was important not to get one's face too close to the sheets when opening the seal, or one could be nearly suffocated by the carbon dioxide, and probably also carbon monoxide and methane which blew out. I did this once, and was left gasping for breath, and could appreciate then how easy it must have been to be asphyxiated at the top of a tower silo, as unfortunately a number of farm workers were, in the early days.

We made some quite good silage, as we mastered the technique, though it was noticeable how quickly secondary fermentation set in once the silo was opened, and it needed to be used quickly if it was not to lose feeding value and become infiltrated with mould. But it was a labour intensive process, and not really worth the candle, so it was abandoned in favour of a proper silo barn, which by then we had accumulated enough funds to build.

In the outside world of silage making, developments followed thick and fast, mostly in the '60s and '70s. Firstly, there was the incorporation of crimping rollers to bruise the grass stems, and liberate the carbohydrates, then the insertion of chopper blades to chop the grass into short lengths, and then the double chop harvesters taking the process a stage further. At much the same time, the use of powder additives arrived to assist fermentation. The powder was fed into the stream of grass as it was blown up the shoot, thus removing the need for spreading it on the clamp. Originally, of course, molasses was the material used to speed up fermentation, and a messy old job it was diluting it from massive drums, and sprinkling it over the clamp with watering cans.

Then, just as one had got used to cutting and blowing and carting all in one operation, came the concept of mowing, followed by wilting in the field. The object was to get rid of excessive moisture and to avoid wastage of nutrients in seepage from the clamp, and also to secure a more reliable fermentation with less risk of butyric acid formation. With the appearance of the modern drum mower, with its tremendous speed and cutting efficiency, and the power of today's harvesters, the pendulum seems to have swung back to a degree of wilting before picking up. And now another pendulum has swung back – to the use of polythene in the field for wrapping and sealing bales of cut grass, which is almost a return to our old vacuum system. If one lives long enough in farming, old practices return, even if they are wrapped up in different clothes – or polythene!

Today, we've reached the point at which more grass is cut

for silage than for hay (not to mention the rapidly increasing acreage of forage maize as well). Huge machines gobble up vast areas of grass in a day, small farmers can no longer afford the machinery and equipment to make it, and contractors rule the world. It is, indeed, a far cry from the days of forking tangled swathes of oats and vetches onto carts, which hadn't even got pneumatic tyres.

The silage story encapsulates much of what has occured elsewhere in farming. The same kind of story could be told about cereal, potato or sugar beet crops, where similar changes have taken place, though perhaps in not quite such a dramatic form.

Grassland Management

It is probably true to say that the appearance of the Wolsey electric fencing unit in the late '30s, had more influence on the intensification of grassland farming than any other single factor. It enabled farmers for the first time to exert strict control over grazing, especially with dairy cows who seem to relish a bite of really fresh grass every day. The concept of controlled grazing dates back once more to the work of scientists such as Martin Jones at Jeallots Hill, George (though not yet 'Sir') Stapledon and William Davies, at Aberystwyth. They studied the effects on the grass plant of defoliation at different stages of growth, and the influence of providing it with recovery periods after grazing. This led in the '50s to the concept of small paddocks which could be grazed down quickly, and then shut up again, fertilised and given a rest of about three weeks before another short, sharp grazing.

This principle was carried to the extreme in one-day-only paddocks, divided by an electric fence, or something more permanent, the size of the paddock being determined by the size of the herd which was to graze it. Personally, I never liked that system, which seemed to me to be unnecessarily complicated and restrictive. It could present problems regarding access to water, excessive pressure of treading, in wet weather, and generally being too small to cut for silage. Instead on our farm we divided our grazing area by the river into three and a half to four acre paddocks, which could be divided into strips each day with a short length of electric fence across them. This was far more flexible and allowed us to alternate grazing with silage making, and to modify the depth of strip allowed each day according to the height of the grass. I also got a licence from Thames Conservancy to extract nine million gallons of water a year for irrigation. This showed some foresight, for when I tried to obtain an increase in the dry year of 1975, we were turned down flat, as they had stopped granting licences by then. The paddocks were a good size for irrigation both with the original Wright Rain sprinklers (which were rather heavy on labour), and for the automatic reel irrigators which succeeded them. Another advantage of that size of paddock was that one could be taken out of the system and reseeded quite easily if it had begun to wear out. We recorded grazing days meticulously, and we could tell quite easily when a particular paddock ceased to pull its weight, which was often after some six or seven years.

With the advent of controlled grazing systems, the use of nitrogen fertiliser, and to a lesser extent phosphate and potash rocketed upwards, with amounts as high as 400 kg of nitrogen per hectare being used. I was never in favour of very high dressings; I had a feeling that the grass produced might not be very good for the cow due to a very low fibre and a high non-protein nitrogen content. In addition, I was not convinced that we would get our money back from very high applications. I recall one afternoon with a class of students trying to do a very rough calculation of how much nitrogen we were getting back in the milk, compared to what we were putting in as fertiliser. We were able to calculate this from our records of gallons of milk produced per acre on the paddocks. We were astonished to find that it worked out at less than a third, even after allowing for the small amount in concentrate feeds that we used at the time. Quite a lot was obviously in the process of being re-cycled through the manure and urine left behind, and a proportion was being temporarily locked up in the soil. It became very noticeable as the years went by how much greener the grass was throughout the winter compared to what it was when we started the paddock system. What was happening to the rest, I don't know, but I suspect that it was seeping down into the subsoil, and thence gradually into the deeper aquifers. No one was interested in the nitrogen content of drainage water at the time, and my concern was definitely economic and not environmental. Perhaps my parsimony then might be rewarded in lower nitrogen content water when it is finally brought back to the surface perhaps 50 years from now. I am fairly sure that the high nitrogen users were on the wrong side of the law of diminishing returns, and I am happy to be in tune with present day thinking on reduced input farming.

Changes in Livestock Farming

Running parallel with grassland intensification was an increase in the number and productivity of the ruminant animals dependent on that grass, especially in the dairy sector. The numbers of cows in the UK increased by about 500,000 over the decade, and average milk yields from 680 to 750 gallons per cow. Herd size also increased. Part of the increased yield was attributable to the change in the breed structure to higher yielding Friesians. Better nutrition also played a significant part, and the economics of production improved through a higher reliance on homegrown grass of good quality. Beef units also increased in number and size, due largely to the production of cross-bred calves in dairy herds. This was made possible by the use of beef type bulls through AI, and the better fleshing quality of the pure-bred Friesian steers fed on good quality grass. Real intensification in grass-fed beef production did not actually come until the following decade, and this was largely true for sheep as well, but the seeds for such intensification were being sown in the '50s through grassland improvement.

It was not only in the ruminant sector that things began to move at an increasing rate, for even more momentous changes were taking place with poultry, both in egg and meat production. Between 1948 and 1960, for example, the percentage of free-range laying flocks dropped from 88% to 28% (by

1975 it was down to 3.5%) and the percentage of battery units went up from 8% to 37% (by 1975 it was 92%). This upheaval in the systems of egg production was made possible by developments in building design and sophisticated environmental controls. High energy mineral and vitamin supplemented rations emerged, as did more effective disease control measures, which enabled large numbers of birds to be kept indoors in cages under very intensive conditions. The payoff to the producer was in much higher food conversion ratios and also increased egg yields and lower labour costs per bird. Genetic improvement played a major part here, as the old pure breeds disappeared and hybridisation became the name of the game. The change was accelerated by the emergence of the large breeding companies, especially in the United States. And it was in America that the revolution in poultry meat production originated as well.

It was, I think, in 1955 that Geoffrey Sykes, a Wiltshire farmer and poultry enthusiast, returned from a visit to America determined to copy the intensive methods of production that had been developed there during and after the Second World War. These involved the mass production of so-called 'broiler chickens'. These were specially bred, very meaty birds, grown to a liveweight of about 3.5–4.4 lb, giving a dressed weight of 2.5–3.5 lb, which took 8–10 weeks to produce. They were grown on in very large houses on litter under the strictest environmental controls of temperature, lighting and ventilation, and fed on very high energy content rations, which provided quite astonishingly low food conversion ratios. They were packed in at very high housing densities of something like 7 lb of bird per square foot of floor space.

There was much controversy about the ethics of the system at the time but it caught on rapidly, and over the next ten years many farmers, large and small, put up broiler houses. But it was, and still is, a system which requires the very highest standards of management if it is not to go disastrously wrong. Many of the earlier producers had problems, due to failure to control disease or to maintain conversion ratios for one reason or another. As a result, there was a tremendous fall-out of the earlier pioneers, and within a few years, a very high proportion of the broiler market was in the hands of a few really large producers, and it has remained so to the present day. Whether one likes the system or not, the fact remains that it has produced chicken for households at a price which is very highly competitive with that of other meats, and its success can be gauged by the 370 million birds reared each year and the average figure of 12 kg of poultry meat consumed per head per year at the present time. A very large proportion of the birds go into deep freeze, and are sold through supermarkets as whole chickens, but over more recent years, quite a significant market has been built up for chicken joints which meet the need for convenience meals for today's small families. It was a truly remarkable development which took place within the space of a few years.

It was not only in the poultry sector that change was accelerating, for in the pig world, things were beginning to move fast as well. It was a time of expansion up to and well beyond the levels of the pre-war industry, and, as elsewhere, it was also a period of intensification. It saw the disappearance of a large number of very small producers and the emergence of large sow herds, geared to the production of either bacon or pork. Breeding sows moved indoors from outside sties or huts, and large farrowing houses with strict environmental controls began to appear, accompanied by dry sow yards, special service pens, and fattening houses. The word 'fattening' was still used then, for it was to be another 15 years or so before the word 'finishing' began to take its place, as leanness became the in-word. Britain was still only producing some 35 per cent of its bacon consumption, as the Danes had quickly re-established their domination after the war, but in pork production, we were over 90 per cent self-sufficient, so there was still room for expansion. The Pig Industry Development Authority (PIDA) had been set up by the Government in the early '50s to build up the market for pigmeat and to improve the quality of the product, and it did good work until it was eventually integrated into the new Meat and Livestock Commission some 20 years later.

The most significant event in the decade was the emergence of the specialist breeding companies, led by that small group of pioneers who formed the Pig Improvement Company in Oxfordshire. This was a co-operative venture, designed to pool the breeding resources of several large producers. Other aims were to provide enough breeding stock for proper genetic testing and a concentration of valuable genes for factors such as leanness, efficiency of food conversion, and so on. The speed of the reproductive cycle in the pig (and the hen, too) makes it an ideal animal for genetic manipulation, but to exploit its advantages to the fullest extent does require large-scale production facilities, which the single traditional pedigree breeder just did not possess. As more breeding companies entered the field, the number of small pedigree breeding herds began to decline — a process which has continued steadily over some 40 years now. As a result of the application of mass breeding methods the shape of today's pig is markedly different from that of the early '50s, and, though not visible to the eye, so too is its conversion of feed into meat. With the emergence of the companies came, too, the virtual disappearance of many of the older localised pig breeds, and a concentration on the Large White and the Swedish Landrace (introduced in the early part of the decade) as the dominant breeds. That situation in fact, changed again about 20 years later as the two main breeds were found to be lacking in certain desirable characteristics, which led to the importation of a number of other breeds to supply perhaps just one or two genes of particular importance.

This era saw a spate of building developments in pig production, mainly designed to make it possible to pack pigs into as small an area as possible, so as to reduce the building cost per pig. It led to the erection of a large number of buildings which proved to be unsatisfactory largely because of faulty ventilation. Many of these were bulldozed down within a few years. I was as guilty as the next man in this respect in putting up a finishing house which turned out to be unsound on ventilation grounds. We were saddled with it for the next 20 years as we couldn't afford to replace it. Gradually, some of the worst problems in pig buildings were sorted out, but by the next decade the notorious pendulum had begun to swing back again towards outdoor breeding.

Significant changes were also taking place in arable farm-ing, particularly in the area of herbicides, fungicides, new varieties of crops, greater use of fertilisers, and mechanisation both in cultivations and in harvesting. But these were more the logical extension of existing practices rather than the more dramatic livestock changes, such as the introduction of broilers, the change to Friesian cows, or the adoption of controlled grazing systems.

STRUCTURAL CHANGE

In the political field, an event of considerable importance has to be recorded – the passing of the 1957 Agriculture Act. Throughout the early '50s overall production was increasing rapidly under the stimulus of the 1947 Act, and the subsidy bill to be met by the Government increased equally rapidly year by year. It became clear that this could not continue, and quite soon there were vociferous critics in Parliament ques-tioning the size of the bill. They asked why farmers, who seemed to be doing very well financially, should receive such preferential treatment compared with other sections of industry. Foremost among the critics was a Midlands industrialist MP, Stanley Evans, who will go down in history for coining the word 'featherbedding', in relation to farmers.

By 1957, it became necessary to put on the brakes, while not abandoning the general principles of the 1947 Act or altering the systems of support. The new Act introduced the principle of standard quantities, whereby the Government undertook to pay the subsidies in full up to a certain tonnage or volume of product, but anything above that standard output would command a much reduced payment. To soften the impact for farmers, the Government undertook to adopt a three year basis for the fixing of prices at the Annual Review, and agreed that the price of a commodity would not be reduced by more than 4 per cent in any one year. This to some extent pacified the farming lobby, as it guaranteed support for some time ahead and allowed farmers to plan for investment without fear of a sudden collapse in prices.

Naturally, the unions made quite a fuss about it, but most thinking farmers realised perfectly well that unlimited subsidies could not be sustained indefinitely, and that some means of curbing them was needed. The new system still meant that efficient operators could make adequate profits, and have something over for reinvestment, while those who could not compete would be gradually squeezed out. The new measures had little effect on production, but did help to raise the overall efficiency of farming.

Changes in the structure of the industry were also beginning to make themselves felt at this time in relation to land tenure. The old landlord/tenant system which had held sway since the enclosures, began to show serious signs of breaking down, with the splitting up of estates, and a change in the attitude of farmers to borrowing money. Up to the Second World War, farmers had been very reluctant borrowers, largely because of the uncertainties of farming, and the fear of being bankrupted if things went wrong and they could not meet the interest charges on their loans. But with the greater post-war feeling of security, farmers became quite enthusiastic borrowers, especially if their landlord – or his heirs after his death – put the estate on the market. This often happened with smaller units on the death of the owner, when land had to be sold to meet estate duty. The logical thing then was for the farmer to borrow from the bank to buy his farm and become an owner-occupier instead of a tenant. It is a process that has continued up to the present day and it has led to a large increase in the indebtedness of farmers to the financial institutions. It is probably no bad thing so long as farming remains profitable. But it could be dangerous if there was a sudden recession brought on by international factors, or unreasonable interference in the normal channels of agricultural trade by Brussels.

THE OXFORD CONFERENCE

While all this was happening in the world outside, on a personal basis the decade was a period of considerable and varied activity, much of it very rewarding and of great interest (to me, at any rate!).

In November 1950 Geoffrey Blackman received a letter from S. J. Wright who was then the agricultural adviser to Ford, but who had been on the staff of the Institute of Agricultural Engineering before the war. The letter said that he and a number of farmers were thinking of reviving the pre-war conference on Mechanisation in Agriculture, which had been held in Oxford from 1936–9. He wondered if the Department would like to become involved as the old School had been. Geoffrey thought it was not quite up his street, but asked me to take it on as he thought we should play some part in it. I was especially keen, as I thought that it might help to get the new Department better known in farming circles. So I attended a meeting at the Farmers' Club towards the end of November 1950, at which several of those who had attended before the war were present. They included Roger North, Clyde Higgs, Reg Older and John Rowsell. It was agreed to hold another conference in January 1951, and a programme was outlined which would look at the changes that had taken place in farm mechanisation since the last conference in 1939. Wright was to do the secretarial work on the programme, and I agreed to do the local organisation, such as the booking of colleges, the Playhouse theatre, and the arrangements for the dinner on the first evening.

The conference went very well, with about 150 attending, and it also paid for itself so there was no need to draw on the £250 that Ford had guaranteed. It was then agreed to hold another conference in 1952, with Wright appointed secretary and myself as assistant. This arrangement operated till the 1954 conference when he retired and I was made secretary. In practice that made no difference, as I had already done all the work for the two previous conferences, when S.J. had taken a back seat.

In 1954, we officially constituted a permanent committee of nine members, from which three had to retire annually so as to ensure fresh blood coming in each year, with myself staying on as secretary/organiser. Thereafter, a new chairman was appointed each year, and the list of chairmen reads rather like a roll call of the personalities prominent in British agriculture at the time.

The conference went from strength to strength, until it outgrew the 500 seats in the Playhouse theatre after the 1956 conference, and we were forced to move to the Gothic splendours of the town hall, with its total seating capacity of well over 1,000. There is little doubt that the conference lost some of its intimate atmosphere in the move, but it did greatly facilitate the serving of teas and coffee, and of the buffet lunches. These were most efficiently provided by a wonderful character, Mrs Mundy, and a band of lady helpers, who really seemed to enjoy the event each year. Mrs Mundy continued to cater for us until she was over 80, and she only gave up when I myself handed over the running of the conference in 1980.

The peak year was in 1963 when the total attendance over the two days was well over 1,000, which was really rather too many. From then on the numbers gradually declined to round about 500–600 in the '70s, due largely, I suspect, to the huge growth in the conference scene about that time, and the emergence of many specialised groups with smaller meetings of their own.

I must not dwell on the conference, since Professor Tony Giles's official history of the Oxford Conference entitled *See you at Oxford* is available, and is a mine of information and comment. I should say that my policy as organiser was always to keep the charges as low as possible, to enable younger farmers and professional people to attend. Administrative costs were low as I did most of the work myself after I had finished tutoring in the evenings. Any official letters were typed by my secretary with the agreement of Geoffrey and his successors. They recognised that the conference was a good public relations exercise for the Department, and, indeed, for the university as well. The popularity of the conference owed a great deal to the agricultural and to the national press, for they always gave us a lot of space and favourable comment. That was the time when nearly all the national dailies had their special agricultural correspondents –

and a very able and distinguished group they were, too.

We only had three committee meetings a year – one in April to post-mortem the previous conference and to settle on a theme and a rough outline of paper headings for the next one; then a second one in June when I presented a draft programme, and speakers were finalised; and a third one in September to add the final touches and discuss administrative arrangements. We generally pencilled in a date for a meeting at Smithfield if it was required, but I only recall one occasion when it was needed, and that was in 1968, when we had to decide whether to cancel the conference because of the dreadful foot-and-mouth epidemic that autumn. This we regretfully had to do, in the light of the Government's requests that meetings of farmers should be restricted as far as possible. Even then, however, the speakers provided the papers that they would have read, which I published so as to keep the sequence of reports whole.

The most time-consuming job as far as I was concerned was the publication of the report, which also included the contributions to the discussions. In the early days this was compiled from shorthand notes made at the time, but as tape recorders became available in the mid-'50s, we used them, sending the tapes to a London firm for transcription. This soon became too expensive, and I found a lady in Oxford who worked from home at about half the price. But editing the transcript was quite another matter, and I don't think that anyone who has not seen a verbatim transcript can appreciate just how different is the spoken word from grammatical text in the English language. Some people, of course, are much better than others in stringing words together into coherent sentences, and in actually being able to finish a sentence before starting another! I must be the first to admit that I came across very badly in this respect myself. Halfway through a sentence, I would suddenly be struck by another idea relating to the topic and go haring after that one – and possibly another one still – before I had finished the first. There were some names that I always dreaded seeing when editing as I knew that I would be in trouble in finding out exactly what they had meant to say. The late Richard

Roadnight, who was one of the traditional 'characters' of the Conference, fell into this category, and I once had to send the transcript back to him, as it was quite impossible to deduce exactly what it was that he was trying to say. I think it must have shaken him, as I had a contrite letter back, and a very short summary. I used to hate précis lessons at school, but, my goodness, I was most grateful to my teachers for their training when it came to writing these reports.

As the number attending increased, it became necessary to establish an efficient front-of-house staff, to take ticket money and deal with queries during the Conference. To start with, Jack Brady, our farm secretary, and my secretary from the Department were able to cope. They were very soon joined by my old student and friend, Geoff Hodgson, who had come back to Oxford to take over the lecturing on crop production (leaving me to concentrate on animal teaching and management). They were then joined by Priscilla Baines, another former student who took two days' holiday from her post as a librarian in the House of Commons to come down and lend a hand. In addition to them, we always mustered a team of current students for holding microphones, taking tickets, and so on. I think, incidentally, that we must have been one of the first conferences to use trailed microphones for the discussions when we introduced them in 1954. Without the help of all the volunteers who gave up their time to ensure that everything ran smoothly, I could not possibly have run the conference properly, or so economically, and I am very grateful for the assistance that they provided. Why they did it year by year, I really don't know, as they were seldom able to get into the conference itself to listen to papers or discussions. Perhaps it was the recognition that they received from what was always an outstandingly friendly audience that returned year by year to renew acquaintances.

For myself, I derived great benefit and much pleasure from organising the conferences. For one thing it brought me into contact with many of the most progressive farmers of the day, in the forefront of agricultural thinking in a period of great technical and economic change. Being by then in charge of our own farm, I found this extremely valuable, and it helped me in no small way to make adequate farm profits, and to provide capital for improvements over a period of 28 years. In that way, the university undoubtedly got its money back from the very small cost incurred by the Department from the use of its services. It brought great pleasure, too, from the large numbers of friends and contacts that I made over the full 30 years of involvement, not just in practical farming but also in the professional and public relations field. Such contacts inevitably led over the next two decades to a number of outside activities of a highly interesting and stimulating nature, some of which continued up to my eventual retirement – and even beyond it.

Finally, I was left with two very tangible rewards, and one intangible one. The first tangible one is the possession of two magnificent silver salvers, the first inscribed, 'Presented at the 21st Conference to Mike Soper with the affection and thanks of the Oxford Farming Conference', and the second marking my retirement as honorary secretary 1950–80. I do particularly appreciate the word 'affection', for it reciprocates so accurately my feelings towards so many of the chairmen and committee members for whom I worked over those years. The other tangible reward was the OBE awarded in 1962, which I have always assumed to have been organised by some of the more influential members of the conference behind the scenes, and whoever they may have been, I am most grateful to them. The intangible reward is a feeling that, as the Conference approaches its 50th year in 1996, and is still regarded as the premier conference and opinion-former in the agricultural calendar, I may have played a very minor part in helping to shape the development of agriculture in the mid 20th century.

There were a number of spin-offs arising from my association with the Conference, which began to materialise about the middle '50s. By then, the numbers of undergraduates entering the course at Oxford had greatly declined (and by the early '60s was at an alarmingly low level) so I did have rather more time available for some participation in outside activities. My first extra-mural involvement was with the Young Farmers' Club movement at national level. I had already had quite a lot to do with clubs at local level, both in Surrey and after coming to Oxford, in the way of speaking at meetings, judging competitions and club efficiency assessment. So I was very familiar with the organisation.

In 1954, Sir James Scott Watson was giving up some of his commitments due to increasing age, and one of these was chairing the Young Farmers Publications Committee at the National Federation in Gower St., and he asked whether I would take it on. I agreed to do so, and was duly elected. This also involved becoming a member of the Council and the Finance and Organisation Committee, and was thereby concerned with all the different aspects of the Federation.

The General Secretary at the time was Ken Savage, and the next few years were of considerable importance to the movement which was in the process of changing from one very much orientated to practical farm crafts, to one which could become a much more generalised youth movement for the countryside. It was already becoming clear that the numbers of young people engaged in practical farming were certain to decline in the future as the industry became more mechanised. It was also very obvious that the membership was already changing quite fast to include large numbers of young people not involved in farming, but who had an interest in some aspect of countryside affairs. It was a time when youth clubs in general were getting a good deal of attention, and funds or assistance in organisation were more available from Local Authority sources. So it was important to show that one was catering for quite a wide cross-section of young people, and not just for an apparently privileged minority of farmers' sons and daughters.

The first task that I was given was to decide whether we could produce a national magazine, free of charge, for every member of the movement – the object being to make them feel that they were part of a national organisation, rather than just of a local club. It was hoped that this would also help to get over the reluctance of some clubs to pay subscriptions to the national body, in which they were not very interested. There were no funds available to pay for a magazine, so the production costs would have to come largely from advertising. There were two snags about this, as we very quickly learned. Firstly, the magazine was going to be given free to the members, and advertisers seem to dislike intensely what they term 'give away' publications. Secondly, the circulation was only going to be about 14,000 copies, which was too small to interest the bigger companies. I think that they also suspected, quite rightly, that a high proportion of the membership was probably too young to have any interest in purchasing the sort of products that would be advertised in the agricultural field. Naturally we tried to sell it on the 'catch them young' principle, but advertisers are not very gullible people and we didn't get very far. We had to employ an agent on commission, which took some of the money away as well, but there was no one in Gower St. who had the time or the knowledge to go touting round potential advertisers.

In the end, we did get enough to launch a journal, and appointed Sue Nathan to edit it, who did an excellent job within the restrictions imposed on her. It got off to quite a good start, but the advertising gradually tailed off, and after a period of some three years we had to close it down for lack of funds. I don't suppose that any Young Farmers of that era even remember it today, which is a depressing thought after the work that was put into it.

The other function of the Committee was the production of a series of very useful farming booklets, produced by Annesley Voysey, and leaflets on club management, and things of that kind. I served on the Council for about eight years, and learned a lot about running a national organisation and the problems of raising money. This was something that I was to become only too familiar with 20 years later when I became Chairman of the Association of Agriculture. I then began to feel that I was getting a bit old to play an active role in a youth movement, and accordingly stood down, but in any case there was insufficient time for it, because of other commitments.

CHANGES IN AGRICULTURAL EDUCATION

Between 1943 and 1973, a seemingly endless series of Government appointed committees were set up to look at different aspects of agricultural education – Loveday, Luxmoore, Lampard-Vachell, de la Warr, Pilkington and Hudson – they sound like a firm of solicitors! The main effect of the earlier reports was a considerable increase in the number of Farm Institutes (later, after the Hudson report in 1973, to be called colleges). They also triggered a growth in the provision of part-time education based both on the Institutes and also on technical colleges and teaching centres run by Local Education Authorities. The aim was to provide for the needs of young people employed in agriculture who could not, or did not want to, give up jobs to go away for a full year's residential course. It was an attempt to bring agriculture into line with the technical training being provided for other industries, where day or block release courses had been established for many years. Up to the Second World War, agriculture had never been regarded as an industry in its own right with a need to provide craft training for its workers – symptomatic, I suppose, of the townsman's widely held view of a farm worker as an illiterate yokel with straw growing out of his ears.

But the war had changed all that, thank goodness, and there was an appreciation in high places that with mechanisation and technical advances rapidly coming into agriculture, there was going to be a need for a properly trained workforce. Oddly enough it was going to be the farmers themselves – or a proportion of them – who were to prove somewhat resistant to the idea, for some years at any rate.

This was where the City and Guilds of London Institute, to give it its full title, came into the picture, as did also a number of regional examining bodies involved locally in technical education. City and Guilds had been set up in 1878 to bring together under a central organisation all the courses and examinations run by the individual City Livery Companies. These dated back in some cases to the 16th century.

By 1938, the Institute's craft and technician awards were recognised by all the major industries as a basic requirement for apprenticeship schemes and very often for pay awards as well. By 1950, it had become the largest single examining body in the world, with over 120,000 candidates a year taking its examinations. Then, encouraged by John Mellars, the HMI at the Ministry of Education responsible for agriculture, it began to investigate the possibility of catering for a number of courses in farm machinery, which had sprung up at various centres across the country. A survey of these was conducted by Sandy Hay and Mrs Olive Foss, of the London Institute staff, who reported that there were enough courses to justify setting up an official examination scheme, which would be popular with the teaching centres.

So in 1953, a small Institute Advisory Committee for Agriculture was set up and our old friend Sir James Scott Watson took the chair. An initial syllabus for machinery operation and care was produced, and the Committee began to look at the possibility of establishing similar courses and examinations in animal and crop husbandry. There was a good deal of doubt, and some opposition at the time to the proposal, on two scores. Firstly, whether farmers would be sufficiently convinced of the value of education and training to their workers to let them attend courses for one day a week for something like 36 weeks in the year. Secondly, whether the workers themselves would be interested enough to want to go to them for any other subject than farm machinery, which most of them did seem very keen to learn about. In the event, syllabuses were produced, but it did take some time for both farmers and workers to be convinced that it was going to be worthwhile. However, by 1960 the principle of day release was well established, and increasing numbers were taking the examinations.

It was at an Advisory Service conference in Oxford in 1956 that I was approached by John Mellars. He had played a leading role in the post-war expansion in agricultural education and in the handing over of responsibility for it from the Ministry of Agriculture to the Department of Education in 1946. The transfer of responsibility had been a very beneficial step, though I remember having doubts about it at the time, as I feared that in some Local Authorities agriculture might be regarded as a poor relation compared to other subjects and be starved of funds. In fact, possibly due to the influence of the HMIs, this did not happen, and on the whole I am sure that far more money was provided by the Department and LEAs than would ever have been available if the responsibility had remained with the Ministry of Agriculture. Clearly, there was quite a difference between LEAs in the enthusiasm with which they viewed agriculture, but on the whole education benefited at all levels from the change of control over funding.

John Mellars asked me whether I might consider becoming an external assessor for a new National Certificate in Agriculture Examinations Board which was being set up to integrate the award of certificates for full-time courses at Farm Institutes. By 1955, there were, I think, 36 Institutes and colleges in the UK, discounting the larger colleges, such as the Royal at Cirencester, Harper Adams and Seale Hayne, and, of course, the Midland and the Wye, which had already been incorporated into Nottingham and London Universities respectively. Institutes were all awarding their own certificates for their one year courses, and there was no central examination system such as the National Diploma in Agriculture which covered the large colleges. The previous year, a few Principals, encouraged by Mellars, had got together to try to formulate a plan for a joint examination, and had actually run a pilot scheme with six colleges. The leading figures had been Don Park in Shropshire, Ken Riley in the East Riding, Ted Stearn in Leicestershire and Peter Marsden in Lincolnshire.

This had worked well, and accordingly a Board had been appointed, composed of representatives of major farming organisations, which was to be chaired, once again, by Sir James Scott Watson. Charles Mercer, a former Principal of Reaseheath, was appointed Chief Assessor, and the City and Guilds had agreed to provide the secretariat, as the RASE did not want to become involved in view of their commitment to the National Diploma in Agriculture. The Institute appointed Olive Foss, a former Oxford graduate, to act as secretary to the Board, and a very good choice that turned out to be, as she was extremely efficient and farsighted, and got the Board off to a very good start.

The basis of the scheme was that each Institute would run its own examinations, and set its own papers according to a given plan, but these would be moderated by an external assessor, who would look after some four colleges, so as to ensure a degree of uniformity. He could insist on changes to the question papers, which would bring them in line with others in his colleges, while the Chief Assessor would hold a watching brief over the whole lot. The scripts were internally marked, but the assessor was responsible for checking the marking standards, and he would then visit each Institute and orally examine every candidate. At the end of the orals, there was an examiners' meeting at which members of the staff were present and the candidates' grading was agreed. There were three grades: pass, credit and distinction, which gave plenty of scope for awarding qualifications to the very varied types of student on the average Institute course at the time.

There were three sections to the examination: section one consisted of papers in crop and animal production, on each of which the candidate had to score 40 per cent to pass; section two consisted of five papers, on machinery, accounts and elementary science – an average of 40 per cent was needed; and section three, was the college's own practical assessment.

I told Mellars that I would be glad to take on the job of assessor, as the scheme seemed an important development which would enable full-time students to obtain a good practical national qualification. The eight newly appointed assessors were a fairly mixed bunch of ex-college Principals, farmers with experience of teaching, two NAAS officers, and myself, and I was the only one from a university. We met together for a briefing meeting in London, at which procedures and standards were discussed, and then we were ready for the road, so to speak. I was allocated four Institutes: Berkshire, Norfolk, Hertfordshire and Leicestershire, in all of which I found the Principals and staff most co-operative and keen to make the scheme a success. It was very hard work getting through the papers, but generally I found the marking by the internal staff very fair, and not markedly different between the four centres, though the standards were not very high in one of them. I found the oralling of the students a fascinating job and quickly learned that the first five minutes often had to be spent in settling them down, as some were so nervous that they could hardly speak intelligently. I always bore in mind how good Jem Sanders and Scott Watson had been when I was in a similar position 20 years before, and that the objective must be to find out how much a candidate knew, and not what he didn't know. I would always make a note when reading scripts of one or two questions to ask a candidate, to check up with them in the oral. It was surprising how often a candidate would say, 'Did I really say that in the paper? Of course, I should have said something else' – and they were usually right! In a way, this also seemed to give the candidate some confidence in oneself as an examiner in actually having read his or her papers.

I acted as an assessor for three years, but then withdrew in 1959, as I had been asked to examine at Cambridge, Nottingham and Wye College, and just could not fit in that as well as the National Certificate. I was not to know then what a part it was going to play in the future. I did leave behind one incident which became part of the folklore of a college – or so I was told some 25 years later. It occured at Brooksby, the Leicestershire Institute, in my second year there, when oralling a student who was very borderline for a pass award. We were out on the farm, and he was the last one before lunch. They had a big new dairy set-up with a long race leading out to paddocks on either side, and this was deep in slurry after a wet summer. Suddenly a thunderstorm crept up behind us, and it began to pour with rain. 'Come on,' I said, 'we had better run for it,' and set off at the double down the race, back to the buildings. Unfortunately, I had previously borrowed some wellingtons from Ted Stearn, the Principal, which were a size too big, and in running back, I tripped and fell flat on my face, diving headlong into the slurry in an effort to save my notebook, which had all the vital information in it. I was absolutely soused from head to foot, some of it even getting into my hair from the splashing. The student turned out to be most resourceful, taking me into the dairy, pointing out the hot water, turning on the hose and finding a towel. While I sluiced myself down (it was the only thing to do, as the slurry was right through to the skin), I sent him to my car to pick up my case, in which fortunately I had a change of clothes. I then crept in to a late lunch, shamefaced but in a reasonably presentable condition. It is probably the only case in history in which a candidate has seen his examiner in his birthday suit! As I have said, the lad was a very borderline candidate on his written work, needing two more marks to be found in two papers if he was to get a pass, and he had not done very well in his oral either, up to the thunderstorm. When it came to his name at the examiner's meeting, I said, 'If you are all in agreement, I would like to find the additional marks for this candidate on his two papers, on the grounds that he showed very considerable initiative in an emergency'. Everyone laughed, and appeared quite happy at the decision, as the general opinion was that he was just about up to standard practically, though not much good on paper.

Other commitments which came my way in the '50s were appointment to the County Agricultural Executive Committee, and to the Drayton Experimental Husbandry Farm Committee. This had started life during the war as the Grassland Research Station under Sir George Stapledon and William Davies, but it wasn't really at all suitable for that purpose as it was on very heavy Warwickshire clay, and with a fairly low rainfall, which in no way was typical of many of the grassland areas of the country. But it was quite suitable as a typical Midlands heavy land farm on which to carry out demonstration work on arable crops, shorter term leys and beef and sheep. So when it was decided to move Grassland Research to Hurley in 1948 (another rather strange choice for that purpose in view of its low rainfall), Drayton quite easily metamorphosed into an Experimental Husbandry Farm. I was to serve on that committee for nearly 25 years, originally under the chairmanship of that great Midlands farming character Maurice Passmore, till I was finally thrown off for old age – probably long after time, too, I expect. But I always found it a particularly interesting committee, and especially so when Ralph Bee was appointed Director, and he began to experiment with minimal cultivations for cereal growing. I was just doing much the same on the farm at Wytham, driven to it, as Ralph was, by the difficulty of getting reasonable seed beds on our respective clay soils in wet years.

Serving on the county Agricultural Executive Committee was also a rewarding experience, made all the more interesting by finding myself on the other side of the table, so to speak, from where I had sat through the War Years in Surrey as a member of staff. We were criticised as being unnecessary in some quarters, but I always felt that we did serve a useful purpose in the years when the Ministry was winding down from the wartime controls, and all the provisions of the 1947 Act were being implemented. We certainly did not hesitate to express opinions on current matters to the Ministry, and to be very critical of policy on occasion, if we considered it unwise.

THE BBC

In addition to these extramural activities, I had become increasingly involved with the BBC from the early '50s. Harry Hunt, who had succeeded John Green at Portland Place, brought me in for occasional contributions on his weekly programme, which had become more of a magazine presentation than straight farming talks, with a number of varied items included each week. At the same time Godfrey Baseley, who was the Midlands BBC Agricultural Producer, commissioned a number of short talks for his programme, which necessitated the odd trip to Birmingham for recording. Godfrey Baseley was, of course, the 'father' of The Archers, and some members of the cast could usually be found in the bar when we finished recording. I would bet that none of them would have believed then that the show would still be running 40 years later! Hilary Phillips, one of our immediate post-war batch of students, whom I had got to know very well, then joined the BBC from *Farmers Weekly*, as a producer. He had journalism in his genes, obviously, as his father was Hubert Phillips who had run the Beachcomber column in the *Daily Express* for many years. His main responsibility was the weekly farming programme, which had lost its evening slot by then, but went out at 12.30 on Wednesday lunchtimes. In 1957, he asked me whether I would like to chair some of these occasionally, which I did in groups of four during university vacations over the next two years.

I remember a sleepless night before the first one, as I had never done an unscripted live broadcast before, nor acted as chairman. It is quite a responsibility introducing a programme and speakers, and controlling the discussions, especially when the other contributors may not have broadcast before and are very nervous. We usually had three speakers in the half hour programme, with a free-for-all discussion after each one had introduced his or her particular topic. We would meet for a rehearsal at 10.30, and it was strange how some programmes would go like a bomb on first run through, and we'd have about an hour for coffee before transmission, whereas on other occasions, things seemed to hang fire, and one just couldn't get them to appear lively. The longer you spent on it, the duller it became. Naturally it would depend on the speakers and their subjects, but even with a good team it sometimes did not 'gel'. It was also odd how one week the live transmission would follow the rehearsal very closely, yet the next week the live programme would take a different course.

The last minutes before 12.30 were always a bit tense, waiting for the green light to go on, and even today, I am immediately transported back into a broadcasting studio, whenever I hear the strains of 'There's no Business like Show Business'. This was the fade-out signature tune of the preceding programme, which meant that in five seconds one would be on the air live.

By the later '50s, black and white television was fully established, and the BBC Agricultural Advisory Committee put in a strong case for a farming programme at a time when most farmers would be able to watch, and they were given a prime lunchtime slot on Sundays. Hilary Phillips was given the job of producing it. He decided on much the same type of magazine format as the radio programme — three or four main items with individual speakers and subsequent short discussions, though one of the live speakers might be replaced by a recording of an event during the week. This would naturally be varied if there were special events such as the Royal Show to report, when much of the programme was recorded. Hilary had to find a strong personality to act as chairman, to whom viewers could relate, and whom farmers would respect as a good practical farmer. He very quickly settled on the late John Cherrington to do this. With his very sharp critical mind, his journalistic experience and his wide knowledge about farming, both at home and overseas, John was a natural for the job, and he did it very well indeed.

But he was quite seriously ill for a few months in 1958, and Hilary had to find some stopgaps as chairmen, and he asked me if I would like to take on half a dozen programmes. I remember that Ted Owens, the well-known Somerset dairy farmer, strong NFU man and one of my past Conference chairmen, was another. I agreed to take it on, though with some misgivings, as I wasn't sure that I would be the right 'type' for television. But it was only for six programmes, and John was expected to be back in due course, anyhow.

The programmes went out from the old Gosta Green studio in the middle of Birmingham, which I think had started life as a cinema. Anyhow, it looked as if it had seen better times, and was in a re-development area, so it wasn't a particularly inspiring location, and it was soon to be replaced by the new headquarters at Pebble Mill.

Live television turned out to be rather more hair-raising than the more intimate feeling that one has in a small radio studio. Somehow, it felt much more exposed on an artificial set, with a lot of open space around one and with the added dimension of cameras recording every movement. As before, we rehearsed at 10.30, and on the first one I got a bad fright (and I suspect that Hilary did, too) when I dried up completely, and just could not think of what I was going to say next. Fortunately he could stop the rehearsal and start the item again, but he insisted thereafter that I must always have some notes on my lap under the edge of the studio table, to which I could refer in an emergency, even if this did infringe the rule that you should not look away from the camera unless you were talking to the other members of the team. I always had a few topic headings as an aide memoire after that, though I don't remember having to use them, except when it came to trailing the programme for the next week, and I was unable to remember the names of the participants. Being an anchorman on a live transmission is really a very responsible job. You have to inspire confidence in your speakers who may be scared stiff, be ready to chip in with a statement or questions as soon as any discussion shows signs of coming to a halt, and you must watch the studio clock like a hawk to see that you are neither over- nor under-running. The former is an unforgivable sin in any form of broadcasting. Then while you are doing all this you have to try to present a lively and confident image, without appearing smug or overconfident. Finally, and perhaps most difficult of all, you have to try to find a way of stopping one of your team meandering on too long, or straying away from the issue in question, without appearing aggressive in cutting them off! This is difficult, as you can't make signs as you can in a sound studio, since the cameras might pick them up.

I was really very nervous on the first programme, and it had not been made any easier by someone saying shortly before transmission, that there were estimated to be five million television sets switched on at Sunday lunchtime. I consoled myself with the thought that they might be switched on, but not many people would actually be watching me dry up on the box!

The first programme went off quite well, with no crises, and two programmes later I had the experience, or perhaps I should say the worry, of introducing the first television programme in which live farm animals were present in a studio. These were six ewes, which Hilary had thought would be a bit of a gimmick, and for which he had erected a hurdled pen in the middle of the studio. Unfortunately, the hurdles were a bit low, and the ewes clearly did not care for the bright lights, and spent most of their time in rehearsal trying to jump out – very nearly succeeding in doing so several times. We were scared that, on transmission, they might succeed, and viewers would have the pleasure of seeing us chasing them round the studio. The late John Cumber, who was a very experienced broadcaster, was demonstrating them, and he was even more concerned about it than I was. At the end of rehearsal, I remember Hilary saying to him, 'Look, John, you'll have to speed up a bit on transmission or we shall over-run'. Of course, the sheep behaved impeccably when the time came for their moment of glory and John, remembering his instructions, whizzed through his commentary at a rate of knots, so that I got a hasty note from Hilary (we were off camera when John was doing his stuff), 'John is under-running by three and a half minutes, you'll have to fill in'. That meant prolonging the final discussion with something we had not covered in rehearsal, and by the time we had just got our teeth into that, I looked up at the clock to find that I only had some 15 seconds left to wind down and trail the following week. This I did most ineptly by getting the names wrong – all because of those damned sheep!

The strange thing about those television programmes was that, after the first one, I was not particularly conscious of any extreme stress at the time, though one was naturally keyed up, but when driving home from Birmingham afterwards, I would quite suddenly feel completely flaked out. On two occasions, I had to drive into the next lay-by and relax for about 20 minutes. I can quite understand how it is that many of those closely connected with live television seem to live on their nerves, and I have never regretted that I did not apparently make the grade as a presenter. I expect that the strain is probably far less today when so many programmes are pre-recorded than in the days when virtually everything was live. Fortunately John Cherrington made a good recovery and chaired the programme for many years, but tragically Hilary died of cancer not long afterwards at a very early age, and I have often wondered whether the stress of the job may have hastened the onset of the disease.

That was not the end of the BBC connection for, in 1963, Tony Parkin, by then in charge of the agricultural sound programmes, invited me to come in on a series of early morning live broadcasts from the farm of Nancibel Gregory in North Oxfordshire. She was a tremendous personality who had played a leading part in local agricultural affairs for many years, as County Chairman of the NFU, and a member of numerous committees as well. I always felt that she was determined to show that a woman could be the equal of a man any day when it came to running a farm. She certainly proved that they could, as she was a most successful farmer, and she stood no nonsense from anyone. Underneath a tough appearance beat a very kind heart. We did six, monthly programmes dealing with different aspects of her farm, and the work that was in progress at the time. We interviewed the men on their different jobs, and Nancibel, both on her farming methods and also on her views about agricultural affairs in general. It was a very popular series, but I don't think that Tony Parkin repeated it anywhere else in that particular format, though his pioneer programmes entitled *Breakfast with* . . . were not too dissimilar.

During the middle '60s I served a five-year stint as a member of the BBC's Agricultural Advisory Committee, and a very interesting commitment that turned out to be. Colour television was then on its way, and I recall a visit to Lime Grove where we had its intricacies explained, and were given a demonstration, which was impressive, though the colour was not yet of the standard that we are used to today. It is difficult to believe that colour has been with us for such a relatively short time – we have got so used to it. Its coming greatly enhanced the quality of farming programmes, in that it made them more watchable for the general public, and thus, to some degree served as a good public relations exercise for the industry. I felt very sad when the BBC axed the Sunday programme, and turned it into one based on the countryside, since it then lost a great deal of its farming content, and became almost another type of nature programme. The old format did tend to provide a counterweight to some of the more extreme anti-farming programmes put out by current affairs producers, which in some cases were unfairly biased against the industry. It has always seemed strange that the producers of that type of programme do not seem to have made use of the experience available to them in the Corporation's own agricultural department.

I would certainly not accuse all current affairs producers of being anti-farming on the basis of a few dubious programmes over the past 20 years, and in fact I had an assignment in the early '80s which proved the opposite when I was invited to prepare background notes and information for the series entitled *All Our Working Lives*. This dealt with the history of 11 different industries over the past 70 years, of which agriculture was one. The producer, Peter Pagnamenta, and his assistants, were extremely fair in their judgements on controversial issues relating to modern farming methods, and the programme when it went out was very well balanced. It provided an excellent historical record of agricultural progress. Looking at the different programmes in that series, it was astonishing to find that there were only about three success stories out of the 11, and agriculture was the most successful of them all. Coal, steel, cotton and shipbuilding were all tragic examples of once great basic industries in terminal decline. It is salutary to think that agriculture might have been included in that list if it had not been for the foresight of the politicians at the end of the Second World War, who formulated the principles that led to the 1947 Act.

RECREATIONAL RELIEF

On a purely personal basis, the '50s was a period of quite intensive recreational activity. On my father's death, my elder sister had kept on the family home in Cornwall, and as petrol came off the ration, it became something of a routine to spend at least one weekend a month down there, generally coinciding with meetings of the committee of the local golf club. This then involved a two-year stint as Captain, and more frequent visits in the summer months for matches and matches and other functions. It often meant late Friday night and Sunday evening driving up and down the old A303, which became extremely familiar. This in turn led to a ten-year spell representing the county on the Council of the English Golf Union. It only entailed two London meetings a year, as I was not an important enough golfer to be put onto special sub-committees, which really did the work. I found it a fascinating contrast to the other national body on which I was sitting at the time – the National Federation of Young Farmers' Clubs, as one could not imagine two more dissimilar organisations. I suppose, in a democratic society, representative councils are necessary for the taking of national decisions, and on occasion they do debate and adjudicate on important issues, but there seems to be an awful lot of time wasted in duplicating work which has already been done. By the mid-'60s anyhow, I decided that I could use my time better elsewhere, particularly as I was by then very heavily involved in agricultural education, which was, after all, more important than golf!

CHAPTER FIVE

The Cereal Revolution – 1960–70

―――᠁―――

MANAGEMENT COMES OF AGE

IF it was the decade between 1950 and 1960 that witnessed the introduction of so many new techniques into farming, it was the next ten years which saw their application on an increasingly significant scale in every sector of the industry. The seeds of expansion sown in the favourable seedbed of the '50s produced plants which, well nourished in the warm, favourable climate of guaranteed prices, grants and subsidies, came to full maturity in the '60s and '70s. Getting bigger became the name of the game – expanded specialised livestock units appeared, heavier yields of crops from larger fields, and bigger machines and streamlined buildings. Chemical products were increasingly used to control disease and reduce weed competition. One effect was a relentless reduction in the manpower on the land – a drop of something like 20,000 full-time employees annually – and a steadily increasing output achieved at lower unit cost. The pressure to 'keep up with the Jones' ' seemed to spur farmers on to ever greater expansion, and inevitably some of those who, for one reason or another, could not stand the pace fell by the wayside. These were often small units which could not produce enough at the prices agreed at the Price Reviews to remain in business. Their land was generally absorbed into the farms of their larger and more successful neighbours, and average farm size steadily increased accordingly.

It was an era when management became the in-word, and this is well illustrated by Tony Giles's book, *See You at Oxford*. He gives a revealing analysis of the topics covered in the annual Oxford Farming Conference programmes over ten year periods from 1951 to 1991. In the 1950s, the themes covered were predominantly connected with technical innovations and their integration into farming practice, whereas in the '60s, the emphasis changed much more towards the principles of business management as applied to farming. Then again, in the '70s, a further change was to occur, with more concentration on the organisation of the industry as a whole and on the marketing of its products and, of course, accession to the EEC.

The movement towards an increasing recognition of the importance of management was given very significant impetus by the publication in 1960 of Keith Dexter and Derek Barber's book, *Farming for Profit*. It outlined the components of business management and applied them to the organisation of production on the farm, and the analysis of costs and returns. Soon afterwards, the concept of gross margins and of fixed and variable costs, first used in Northern Ireland, and then perfected by David Wallace at Cambridge, was introduced. This certainly helped farmers and their advisers, who at the time were mainly NAAS officers, to analyse their costs of production more effectively. For a time this had the effect of making many people think that a gross margin figure for a crop or livestock enterprise was the be-all and end-all of the exercise. But it was not long before it was recognised that a high gross margin figure did not necessarily mean more money in the bank at the end of the day, and that it was the net farm income that was the more important measure. But the concept has stood the test of time, and it has undoubtedly helped many farmers to streamline the management of their farms, and to eliminate enterprises that were not paying their way.

It certainly came as something of a shock to some farmers to find that some of their pet enterprises (a new word, incidentally, that came into popular use at about this time) were, in fact, losing them money. The new method of assessment had the disadvantage that it tended to break the farm down into specific compartments, with gross margins for each, and it was sometimes difficult to pull out the interactive effects of one enterprise on another. I always felt, for example, that the average smallish sheep flock on a mixed farm came off badly in this respect, in that it generally achieved a low gross margin, but it could have a beneficial impact on other enterprises which did not show in the figures. There is little doubt that the introduction of management principles did lead to a more rapid acceleration in the move towards specialisation, which was such a characteristic of the period, and probably pushed it too far in some cases. This is best illustrated perhaps by the growth in the number of specialist cereal growing farms which led to the creation of very large fields in some areas, and to the indiscriminate ripping out of hedges in the name of efficient mechanisation management. It also led to the coining of emotive words, such as 'prairies' and 'barley barons' which have haunted the farming industry ever since, even though the continuous barley grower disappeared for the most part long ago, and today there are more hedges being planted than taken out.

I don't think that it was fair to blame the gross margin system as such for excessive specialisation, which some people have done, for the trend towards the elimination of small subsidiary farm enterprises had started long before Dexter and Barber's book saw the light of day. Our most successful Oxford Conference in the early '50s was entitled 'Specialisation or Diversification?', and the consensus from it came down very firmly on specialisation, mainly on the grounds of the more efficient use of labour and machinery. It was argued, I recall, that the old dictum about the dangers of having all one's eggs in one basket was out-dated in an era when prices were guaranteed at a certain level for some four years ahead. This was true at the time, though it took no account of other factors, such as adverse weather conditions, disease outbreaks or other Acts of God which can prove to be potent reasons for an 'egg basket' theory. It was also argued that these were only rare occurrences, and that the saving in costs year in, year out, would counterbalance an occasional disaster, and that belief was to hold sway for many years to come.

STRUCTURAL CHANGE IN THE '60s

The sort of changes that were taking place in the industry can be well illustrated, though on a very modest scale, by what was happening on our own farm at Oxford.

It had quickly become apparent that, in the light of the speed of technical change, we had planned things on far too small a scale, even for a 300 acre farm. The milking parlour was too small, and difficult for the herdsman, as were the calf boxes and youngstock pens, as they all had to be mucked out by hand. The pig unit at 18–20 sows was ridiculously small and the pens were labour intensive. The tractors were too low-powered to get good seedbeds on the heavy land quickly enough if the weather was at all tricky, and handling grain in sacks off the combine and onto the sack drier was also very labour intensive. The need for the drier for experimental plots disappeared when the Weed Research Organisation was set up in 1960, and part of Geoffrey's team moved over to its new farm at Begbroke to the north of the city. It was quite impossible to try to rectify all these things at once, as we were dependent on being able to make enough money for probably only one major improvement for expansion annually. Fortunately, by then we had pulled the arable land round, and yields had greatly improved, and the dairy herd was making money, so the University Chest was quite happy to let us go ahead with a new project each year.

Herringbone parlours had by now been introduced from New Zealand, and one of the first things was to adapt our old parlour to a 5 × 5 herringbone (and how ridiculously small *that* seems today!). We also wanted to up the size of the herd, initially to 65 cows which we could still house in the original straw yards by stripping out the feeding passage. But this meant providing facilities for bulk-feeding silage. The logical step was a silage barn with sleeper sides for self-feeding (which by then had proved to be a satisfactory system) and room for storing straw for the yards on top of the clamp. We then found that with self-feeding, there were always cows outside the yards or in the hay feeding passage, so that we could increase the cow numbers still further without too much pressure in the yards themselves. So with the new parlour and silage set-up the herd size crept up to about 80, though in practice that was probably too many. To offset that advantage the yards then had to be mucked out twice a year, which was inconvenient, but a small price to pay for the larger herd.

We had been forced to increase the herd size rather more than I had intended by the retirement of a neighbour, which led to the university very sensibly deciding to split that farm between ourselves and our other neighbour. It brought another 120 acres, comprising about 30 acres of light arable land, and another 80 acres of grassland adjacent to the river. That meant more stock to graze it — largely youngstock, as much of it was out of reach of the dairy herd. I preferred to do this by natural growth of the herd, though we did buy in steers for two years till the gap could be filled with our own cattle.

The next priority was to cut the labour at harvest. It was an economic drain because there was both a bigger cereal acreage to deal with, and yields had increased quite significantly. This was the time when moist grain storage in sealed silos was being perfected. I thought that this ought to suit us well in view of my policy of home mixed food for the livestock units, and particularly as Dr Reg Preston's system of so-called 'barley beef' had recently hit the headlines. I wanted to give that system a trial, even though it did mean diverting some of our steers away from the grass. So we put up a 90 ton moist barley silo and scrapped the old sack drier in favour of a smallish Alvan Blanch continuous flow drier. This necessitated completely reorganising the grain handling system in the barn, with additional dry grain bins for which fortunately there was still room. On the whole, the moist grain silo, which we filled with barley each year, worked pretty well, and the barley rolled very well for the cows and the barley beef unit. It all cost a lot of money, but it reduced the grain handling at the farm to a one man unit. The combination of wet and dry storage in particular gave us much greater flexibility at harvest as we were not so dependent on the throughput of the drier. But the whole exercise does illustrate very well how a change in one piece of equipment can necessitate further changes elsewhere.

In the mid '50s, Massey Ferguson had introduced a machinery rental system which was available to colleges, whereby tractors, combines, drills, etc, could be hired on an annual basis with an option to purchase at the end of three years. This suited us well, as it meant that we could get the use of new equipment without having to use capital urgently needed for other improvements. As we didn't pay tax, we could not benefit from tax allowances on new machines, as ordinary farmers could, so renting was an attractive proposition.

Initially we rented two tractors: one large and one medium model, a combine and a grain drill. The larger tractor made a big difference to the cultivation of the heavy land, enabling us to get better seedbeds, and crop establishment. We continued to rent tractors and usually bought one in as the three years expired, until we got to the point of having more tractors on the farm than men to drive them. This can be an advantage provided that they are well written down, as the older ones can be used for special purposes with appropriate equipment kept on them, thus saving a great deal of time and labour continuously changing it.

By the early '70s, I found that the rental charge for the combine had got too high in relation to the acreage of corn that we had on the farm, so we stopped renting and bought a second-or even third-hand one about every three years. This was at a time when the big arable farmers round us traded in their combines – largely for tax reasons, I suspect – after about two to three years, so there were generally reliable ones to be had secondhand.

After the reorganisation of the grain harvesting, the next vital thing was to get a better grip on the pig unit which was overcrowded and not properly equipped with buildings. The first step was a proper farrowing house, so that we could dispense with the outside farrowing huts, and bring the sows indoors, except for dry sows running out in an old gravel pit in the summer. This would also allow us to double the numbers. So a new farrowing house was built and I kept my fingers crossed that it would prove to be more successful than the finishing house with poor ventilation with which we had saddled ourselves in the early days. At the time it was still difficult to get reliable advice on new livestock buildings because there were so many new ideas floating around. The more advice one sought, the more answers one was given, and the more difficult it became to make a decision, since few of the new buildings that one could inspect on other farms had yet stood the test of time. It was especially difficult with pigs, with so many factors to be considered: not only ventilation and warmth, but also problems of dung removal, piglet safety, sow welfare, working conditions for staff and labour-saving devices. We finally settled on a composite design for a 20 sow house with built-in crates incorporating ideas gleaned from several sources. It worked quite well, though it was a bit heavy on labour. It was, at least, a great deal better than the finishing house.

In that connection, incidentally, we were visited by Professor Brambell and one or two members gathering evidence for his Committee on Animal Welfare in 1964. Their comment about the house was that they had seen worse, a comment that I took to mean that they, quite rightly, thought it was pretty bad, but were too polite to say so! It was the Brambell Report in 1965 that advised against compulsory legislation in regard to animal welfare, but made a strong recommendation for the establishment of codes of practice instead. I felt at the time that this was a sensible decision since I could not see how any legislation worthy of the name could be properly enforced, and there was still insufficient knowledge about buildings to be dogmatic enough to formulate rules and regulations.

So, by the mid '60s, 15 years after we had taken over the farm, every building that I had put up or modified at the start had been drastically altered or added to, with the exception of the machinery shed and workshop which we had actually planned on a large enough scale. That illustrates very well the pace of change at the time, particularly in respect to the expansion in the size of animal units, and to the realisation of the importance of labour-saving devices. Even at that time labour costs were steadily rising year by year.

Those changes at home mirrored what was happening, and what has continued to happen up to the present time, in the farming world outside. Between 1960 and 1990 the size of the average dairy herd increased in the UK from 38 to 62 cows, while the number of milk producers fell from 160,000 to only 48,000 – and it is now lower still – an astonishing drop in numbers. Over the same period, the average yield per cow increased from some 3,500 litres to just on 5,000 litres (770 to 1,100 gallons) per lactation.

Almost exactly the same picture can be seen in the pig industry, where the number of breeding herds has declined over the same period from 79,600 to only 12,000. Simultaneously, the size of individual herds has increased from 10 sows to 62, and even that figure seems low, since 75 per cent of breeding sows are now in herds of over 100. These figures represent very clearly the remarkable transformation in the management of those types of farm animal which require a relatively high capital investment in equipment and services. There generally comes a time on smaller farms when buildings and equipment become obsolete, or market pressures become too great. A decision then has to be taken whether to invest capital in improvements or expansion, or to get out altogether. In many cases, as the figures show, the decision is to get out, and either try something else or sell up.

It was in the '50s that the late Richard Roadnight developed a system for keeping breeding sows out of doors on the light land of his Oxfordshire farm, using a crossbred sow derived from the hardy Saddleback sow mated to a Landrace boar. This produced a good mothering sow capable of standing up to tough winter conditions, and to farrowing in portable huts. The system steadily caught on during the '60s, and finally led to the appearance of very large sow herds kept under these conditions. With the extension of the animal welfare movement, the number of such herds is now very large, accounting partly for the great increase in average herd size.

In the beef world, there was a fairly wide adoption of Reg Preston's system of intensive concentrate feeding during the '60s and this coincided with the human change in eating habits, which had started after the war. There was a swing away from larger joints towards small cuts suitable for just one or two meals for the reduced size of family, and a trend towards convenience foods. The 400–425 kg carcase produced by his system, linked to the fact that the meat was very lean, suited this market admirably. Modifications of the system quickly appeared, including feeding varying amounts of silage or intensive grazing to produce small carcases at 18 months of age, instead of the 10–12 months of the Preston system.

We did adopt the system on the farm, using our own rolled moist barley at 85% of the ration and 10% soya bean meal, but I also added 5% of white fish meal, in addition to a double dose of vitamin A supplement in the mineral/vitamin component. I had read somewhere that vitamin A inclusion helped to avoid the possible liver damage which was liable to occur in a few animals when they were fed high levels of barley. I was also indoctrinated with the virtues of white fish meal at Cambridge, where we were taught that it was a very useful supplement in the rations of ruminants, even though most people associated its benefits with just pigs and poultry. It wasn't until I appreciated that in feeding the ruminant, one is not really feeding the animal directly, but only indirectly through the micro-organisms in the rumen, that I realised it was they that the fish meal was benefiting in the first place. These two additions to the basic ration increased the cost, but the inclusion of the fish meal reduced our average time to slaughter by ten days. We only ever had one case of liver damage over the eight years that we ran the system, and practically none of the foot problems, which occurred in some units, I believe.

Incidentally we produced these cattle for several years in a makeshift yard, roofed with polythene. The uprights were old telephone poles that could then be bought quite cheaply, and the timber-frame roof was of very light construction, as it was bearing virtually no weight. This was constructed with the help of Ron Spice who had previously helped us with the vacuum silage. The first roof tore very quickly within two years, but the second one we covered with netting which kept the polythene in place. This lasted longer, and provided very cheap yarding, but the system never caught on in practice at the time as the material was of too thin a gauge. The principle has, of course, been adopted since then for the provision of relatively cheap winter housing for sheep.

We finally gave up intensive beef, not because it was unprofitable (although it was becoming less so) but because we got another addition to our acreage with the death of our second neighbouring farmer. This gave us more low-lying pasture, which I decided to use for extensive beef production, which I deemed to be a better use of our resources.

Our gross margins were always very borderline using the extensive system, but that is one of the puzzling aspects of beef production. No economist has ever really made beef look financially attractive on paper, and yet thousands of farmers continue to produce it by a variety of methods and still seem to stay in business.

I suppose that I might have introduced a single suckler herd instead, but I could never see the economics of a system in which a sizeable cow is fed for a whole year just to produce one calf and supply it with a few hundred gallons of milk. If the amount of energy contained in the dry matter of the food taken in over the year is balanced against the amount of energy in the dry matter of the calf at birth and the dry matter of the milk it drinks, the equation looks crazy. I remember reading once that the efficiency of a railway engine in the days of steam was of the order of 12 per cent, but that of the suckler cow must rate far below that! The system can only pay, surely, if the rental value of the land and the cost of the feed is extremely low or the resulting calf is of exceptionally high value.

In the early 1960s, it was sheep that seemed to be lagging behind in the race for expansion. This was partly because on many lowland farms they were often very much a secondary enterprise as far as animal production was concerned. The situation was not improved by low prices due to the very high level of imports from New Zealand with its lower costs of production and efficient marketing system. In 1960, for example, out of a total national consumption of mutton and lamb of 600,000 tons, Britain only contributed 240,000, or 40 per cent. There were, it is true, pioneers coming on the scene who were adapting sheep production to the more intensive systems of grassland management established for dairy herds. But to cover the extra costs involved, it was essential for such farmers to increase their output. That was the problem, for there are, basically, only two ways of intensifying the sheep enterprise — to produce more lambs per ewe or to keep more ewes per acre. The former was not easy, due to the very large number of breeds, and the fact that prolificacy is not of very high heritability so that it can take a fair time to concentrate the relevant genes. It was surprising that breeders had not concentrated on this before. But the blame lies at the door of earlier pedigree breeders whose object in life seems to have been to win prizes at shows for what were considered very beautiful animals, regardless of their real productive qualities.

But the pioneers like Oscar Colburn, Henry Fell, and Stephen Hart, to name just three, were prepared to discard such irrelevancies, and adopt the sort of methods which had started to transform the breeding of pigs in the previous decade. As a result of selection within breeds, the importation of new breeds and crossbreeding, prolificacy began to improve quite markedly through the '60s. It was, however, to take about another 20 years before 200 per cent lambing flocks became more or less commonplace.

The other aspect of increasing output per ewe was to improve survivability in the lambs, which in many flocks was low, with as many as 20 per cent of lambs born alive being lost. Husbandry methods improved greatly over the next 20 years to improve this figure, though it became more difficult as the percentage of twins and triplets increased, because of smaller size at birth.

Perhaps the more immediate objective of the average

farmer, though, was to keep more ewes to the acre of grassland, while not at the same time decreasing output per ewe through overstocking. That meant higher applications of fertiliser and controlled rotational grazing systems to maximise grass growth. This helped to reduce parasitic infestations, which get out of hand so easily at high rates of stocking. This did happen in the early days at rates of 7–8 ewes to the acre. By the end of the decade, one heard talk of housing ewes during the winter in order to keep them off the grass, so that it could start earlier and carry more stock in the spring. After a late start compared with other types of livestock, sheep production was definitely beginning to move into the modern world by the beginning of the 1970s.

CHANGES IN ARABLE CROPPING

In arable farming the decade of the '60s was a time of intense activity, especially in the cereal sector with, possibly, the explosion in the barley acreage being the most dramatic change. In 1946, it had stood at 2.2 million acres nationally, which increased to 3.4 million by 1960. By 1970, it was 5.5 million after having peaked at no less than 6.1 million acres in 1966, a figure which will probably never be exceeded. A proportion of this was at the expense of oats, which lost a million acres over the decade. But much of the increase was due to two other factors: firstly, the emergence of new varieties with winter hardiness, stiff strong straw and higher yield potential; and secondly, a realisation that barley was quite a suitable grain to feed to ruminants in moderation. Another factor in its favour was the availability of new fungicides which made it possible to grow the crop continuously without too much risk.

This was nothing new, for the late Fred Chamberlain had done it for at least 20 years on his farm at Crowmarsh, near Wallingford, between the wars. His yields naturally were much smaller than those which were obtainable in the '60s, and he relied on undersowing with rye grass and trefoil to maintain soil structure.

The derelict farmyard at Wytham on take-over in 1950

The same after the first building renovation

Vacuum silage, which was experimented with for three years at Oxford. The clamp after extraction of air in the evening

Early days making clamp silage. This illustrates the risks which were taken before the clamp silage was developed. Ted Floyd, shown in the photograph with his sons Ted and Mick, worked on the Oxford farm for over fifty years.

The Oxford farm slurry separator with reservoir in the background and slatted collecting channels from the kennels and other buildings in the left foreground

A frustrated author vainly trying to filter separated slurry effluent through straw bales. The newly excavated one million gallon reservoir is seen in the background

My dear friend Joan Bostock MBE FRAgS, flanked by Lord Henry Plumb and Charles Jarvis, following a presentation on her retirement as General Secretary of the Association of Agriculture after 38 years service

A familiar sight to generations of students during their tutorials

Aspects of the situation in the '60s compared with the previous decades illustrate how opinions can change in farming in a relatively short space of time. I was taught in the '30s that wheat should only be fed to poultry, that barley should be fed to pigs and certainly not to ruminant animals, and that oats were the best and only safe cereal for ruminants. But what do we find in the '60s? Wheat being fed to pigs at quite high levels, barley being very widely used for cattle as well as wheat, and oats being used very little except for horses.

Wheat did not take off at anything like the same rate as barley, for the total acreage increased by only half a million acres - from 2.0 to 2.5 - during the decade. This was probably because it was still regarded as essentially a crop which required strong land if it was to yield well. It also needed to be grown in a fairly wide rotation if it was not to suffer severely from fungal infections such as take-all and eyespot, for which no reliable treatments were available. A problem with winter wheat was grass weeds such as Black Grass and wild oats, which proliferate rapidly if autumn sown crops are included too often in the rotation. These were the main reasons why some of the early pioneers of mechanised arable farming, experimenting with continuous cereals in the '30s, came unstuck, and went back to including leys in the rotation.

It was a problem that we had to face on our own farm, where much of the arable land was on heavy Oxford clay. I didn't want to have to grow leys since we already had enough grass on the wetter fields by the Thames for the amount of stock that we could carry, unless we went into sheep. I was not keen to do this for several reasons: our fences were nonexistent on the arable block, the low-lying land had a liverfluke problem which we had already encountered with the cattle, and it was likely to be bad for footrot as well. Furthermore, we had nobody on the staff experienced with sheep, and at the time they were not, as I have already said, attractive financially.

There were not many alternative crops to wheat for our type of heavy land. We grew a field of beans each year, but it was a risky crop, and I was not prepared to put aside a big acreage for it, as the market was very limited then. In the early days when we were trying to build up fertility and clean the land we did grow leys and farrowed the sows on them, but we had a couple of wet winters when they got in an awful mess, and it really wasn't on as a long-term solution. The two main crops which Geoffrey Blackman was investigating were oilseed rape and forage maize, and we had a lot of plots of both on the farm, and they seemed possibilities as break crops. The problem with oilseed rape then was not so much in the growing of it, but what to do with the seed when it came to marketing, since the main crushers could get ample supplies of oilseeds from overseas and were just not interested in small quantities of homegrown rapeseed. It was going to need a very big change in world markets which did not come until the economic upheavals of the mid '70s, and the emergence of a British marketing organisation, before oilseed rape could become established as a reliable and viable crop. Sadly, Geoffrey Blackman, who had great faith in the crop's potential, did not live to see it established.

I was rather more interested in forage maize for silage, possibly because I had seen it grown as green forage on both the Cambridge and Reading farms for the dairy cows in a dry August/September. This was the old white Horse Tooth variety, which was very strong growing, but too late maturing to produce cobs, and therefore useless for silage.

We grew some of Blackman's newer varieties for several years for silage on medium to heavy fields, until one year we had a very wet autumn and got into a fearful mess trying to get it off the field, with trailers and tractors getting bogged down to their axles. I decided then that it was not a very feasible break crop for our land with the primitive tackle then available. Blackman had got one of the machinery firms interested in developing a harvesting attachment to fit onto the front of a conventional flail grass harvester. This took one row at a time, which was fed in through guide bars set about a foot above the ground, from where the stems were turned down and fed into the harvester. It was fine so long as the crop was standing upright, but if it was leaning over at all – as it often was – it became a tedious and frustrating job getting it into the harvester at the right angle. This was in the days before chop mechanisms had been incorporated into forage harvesters, and the flails lacerated the crop into uneven lengths, which made it very difficult to get adequate consolidation in the clamp. We lost a great deal of energy through carbohydrate breakdown, especially when the clamp was opened up and the air could get in.

I did a simple trial for two years in the early '60s, comparing milk yields from seven paired cows – half of which were fed grass silage and the other half a mixture of maize and grass silage. The yields were comparable for the two groups in the first year, but in the second, when the grass silage was of rather poor quality, the combined maize and grass certainly came out best, with far less day to day variations in yield. Grass silage in those days, even in the early '60s, tended to be more variable throughout the clamp than it is today, since it could not be made quickly on the average sized farm. This meant that weather changes, delays in cutting, grass coming in from different paddocks and so on, all made a difference to its uniformity in the clamp.

Looking back, after 30 years, at our experiences with maize and oilseed rape, I feel perhaps that we should have soldiered on with both crops in the light of what has happened to them since. Both are now fully established as major farm crops, but their success has depended entirely on developments outside the farm gate. In the case of rape, it has been the willingness of big companies to process the crop, due to the changes in world market conditions. To some extent the introduction of new varieties with better chemical properties, and of more suitable herbicides also helped. In the case of forage maize, it has been the development of high-powered tractors, and sophisticated harvesters capable of chopping very large amounts of crop into uniform segments at great speed, which has made it a physical and economic proposition. This once again endorses the theme that keeps recurring over these 60 years of farming progress – that the application in farming practice of new knowledge gained from research may have to wait for many years for developments in other fields, before it can be fully exploited.

It had become very clear as soon as we had begun to get the arable land on the farm into a reasonably productive state, that we had two problems if we wanted to grow a high proportion of cereals on the heavy land. The first one was to get good enough seedbeds, especially in the spring, and the second was the presence of grass weeds. In a way the two were linked, because we had to grow a proportion of spring cereals in trying to control the grass weeds. Ideally, on that sort of land, one would have liked it to be entirely planted up before the winter, so that if it was wet, there would be no problem of late sowing in the spring.

HAVING FUN WITH DRAINS

One of the first things when we took over was, of course, to get the ditches dug out. Some of them were virtually non-existent and overgrown. We found clay pipe outfalls here and there, but most of them were very deep and narrow bore, probably dating back to the 1850s or '60s. They were all silted up and ineffective and it was clear that a great deal of money would need to be spent on proper underdrainage at some time in the future. But the money wasn't there then, particularly as we were not allowed the 50 per cent drainage grants available at the time. By the early 1960s when we had finished some of the rebuilding work, I felt that we just had to tackle the drainage problem, and from then on tried to set aside a few thousand pounds each year to work round the farm, starting with the wettest fields. Then, as soon as we had done that, we got an addition to the acreage and had to start all over again.

Plastic piping had just come in, and our contractor was very keen to find farmers prepared to give it a try. I was more than willing to do this, as I thought that it must be more sensible than lugging great weights of small clay tiles around, and spending time ensuring that they were really properly laid. Where the layout of the field permitted, we adopted a system of fairly widely spaced plastic laterals feeding into six inch tile mains, which themselves fed into the ditches at strategic outfall points. All drains had 12–18 inches of shingle over them, and initially we mole drained across the laterals. Thereafter we subsoiled at about three year intervals, which became necessary in the light of a later decision to adopt tined cultivation. I am not convinced that plastic piping is really as effective as old-fashioned tiles, as it must have a shorter life span, but it probably wins out in terms of cost and ease of installation. I was always fascinated by land drainage, influenced by that delightful character, H. H. Nicholson, who taught me at Cambridge, and who was responsible for some very good work on mole drainage in the late '20s. There was little that I enjoyed more than walking ditches in wellingtons after heavy rain and seeing the water trickling out of newly installed pipes – or better still, spouting out of them. Of course, it was not always quite such fun, as it is infuriating, after spending a lot of money on a scheme, to find that some of the outfalls are not running, and one wonders what may have gone wrong.

When we took over the land from the second farm in 1970, there was obviously a major drainage problem in the middle of one of the large arable fields, where a virtual bog of about half an acre in extent had formed. It had been increasing in size year by year as the previous tenant found it more impossible to get through it. Some years before, someone had introduced me to water divining, using metal rods (I had tried the conventional hazel twigs before without success, but the rods worked like a charm). It is quite astonishing how they will reveal the location of any underground pipe or water source. For those not familiar with the technique, you take two thin metal rods – welding rods are very suitable – and simply bend about three inches at the end of each rod at right angles. Then hold them loosely in the hands with the short bend between the thumb and the palm of the hand with the longer part of the rod horizontal to the ground and pointing out directly in front of you. As you cross a drain or pipe, the rods swing together across your stomach. It is quite uncanny, for it works even if the drain is under deep concrete. What the hidden force is that diverts the rods, I have been unable to discover, but it seems almost infallible with those for whom it works. I used to try it out with students on farm classes, and it was effective with most of them, but there were a few who got no reaction, which makes it even more puzzling. I was intrigued to find, when gas was being laid on in my village recently, that the foreman of the gang used his rods to locate underground water and sewage pipes in the road, and he claimed that it was 100 per cent effective. Sometimes, one could get a strong reaction, but on digging down could find no trace of a drain. Possibly they were there but very deep, as there was a school of thought in the 1860s that they should be put down at a depth of three feet.

I had a fine time on the new farm, tracking all the drains leading down to the boggy area, and sure enough when we dug down at the point where three appeared to converge, we uncovered quite a large old stone drain, probably laid soon after the enclosures, and before clay pipes became available in the 1820s. Not surprisingly, this was completely blocked up. There was no point in trying to restore it, and as it was at a good depth, we were able to lay a completely new system over the top of it. This worked well, and in a couple of years there was no trace of the previous wet area at all.

NEW IDEAS ON CULTIVATION

In the early '60s, I had begun to revolt against the sight of our glazed wet clay squeezed up by the plough. It was very difficult, if not impossible, to get this worked down to a decent state of tilth, unless it had been exposed to some good sharp frosts, or to some alternating periods of wetting and drying. It was still partly a legacy from the complete lack of organic matter that we had inherited. It was also a characteristic of our particular brand of clay which contained quite a high proportion of fine silt particles as well as the clay. It seemed to me that it must be wrong to deep plough the organic residues we were getting from the improved crops that we were growing, and that the sensible thing would be to leave these near the surface where they would improve the topsoil structure and help seedbed formation. I then recalled three things. The first was a book called *Ploughman's Folly* written by an American, called Faulkner. His argument was that the plough was one of the worst implements devised by man, as it put down organic matter from the crop and brought up poorer soil instead. He was, in fact, writing about completely different conditions to ours, and was concerned about the breakdown of organic matter through aeration on dry land, but the principle seemed relevant. Secondly, I remembered the very heavy clay soils in the Sudan Gezira, where we never turned the soil over (to avoid drying out) but cultivated it and kept the previous crop roots near the surface where they greatly improved the soil structure. My third recollection was that of seeing steam tackle at work on the heavy land at Cambridge in 1935, which was worked with a heavy cultivator.

While I was pondering this problem, the first so-called chisel plough came on the market. This wasn't a plough at all, but a very heavy cultivator with chisel tines. It seemed that it might be the right tool to stir up the surface soil into lumps after harvest, which we could then break down more easily than a furrow slice, because the soil would be more friable. We bought one straightaway, and though it was not ideal — machines seldom are for all types of farm conditions — it certainly went quite a long way to doing what I wanted. Its disadvantages were that if the land was very hard and compacted after harvest, it might bring up huge lumps of earth the size of one's head; or conversely if the soil was very wet on top, it cut up slivers of putty-like soil. But if the land was in a reasonable condition, it did quite a good job, though it was slow work, especially with the relatively low-powered tractors then available.

By coincidence, at the Drayton Experimental Husbandry Farm for which I was a member of the Committee, we had just had a change of Director, with Ralph Bee succeeding Stanley Culpin. Ralph had the same feelings that I had about bringing up wet clay with a plough, and then trying to work it down again in a tricky season. He moved over to surface cultivation in place of ploughing, too, so I was able to keep a watching brief as to how he got on. To cut a long story short, we seemed to get on well on the few fields that I was brave enough to experiment with at the start, and within three years we had stopped ploughing on the farm altogether, unless it was for a special purpose such as breaking up a ley for reseeding. Even for this in the 1970s, we went over to direct drilling on a number of occasions. At about the same time that I adopted a no-ploughing policy, I decided that I would also start experimenting with continuous cereals, to see if, with the herbicides and fungicides by then available, we could get reasonable yields. To begin with, in 1964, we set aside one field for continuous wheat on the heavy land, and one for continuous barley on the light land. On other fields we rang the changes between winter wheat and spring barley in no particular rotation, with one field of beans each year.

I was too cowardly to commit more than a seven acre field to the continuous wheat, in case some horrible disaster hit us. It didn't, and 18 years later when I retired, it was still growing wheat, with much higher yields than when we started. We were nearly beaten after five years by wild oats, and would have been, if it had not been for the fortunate arrival on the scene of Suffix. I managed to get enough of this to spray the field the year before it was launched, through the good offices of Roger Pierce, who was then the Farm Director at the Shell Research Farm at Ilmer. It did a very good job for us, and thereafter, though there were usually a few plants visible, wild oats did not trouble us again seriously.

The wild oats were succeeded before long by a far more insidious invader in the form of sterile or false brome, of which we first really became aware, I think, in 1970. Then Black Grass, from which we had been completely free, suddenly put in an appearance as well. It was limited at first, but built up quite fast on the continuous wheat field, as might have been expected, since herbicides for its control were not as widely available as they are today. But in 1969, I had recalled how, in the Sudan, we burned every bit of cotton plant residue after harvest. I thought that this might work for weed seeds, as well as for disease control. So that year, after harvest, we burned the stubble on the continuous wheat field. I don't claim that we were the first to have started the practice but we were certainly one of the first farms to do it in the area. After two years, we spread the straw and burned the lot. I was very pleased with the result, for it left a completely clean, scorched surface for the cultivator to break up.

One year, I'm afraid we got too close to a hedge, and damaged it, which made us more careful in the future to turn the straw well away from the hedge before putting a match to it. This was before the days of codes of practice, of course. Under these, the introduction of headland margins, cultivated instead of being burned, in the end became our undoing. This did not destroy the grass weed seeds, but more likely gave them a nice seedbed, and the sterile brome began to build up rapidly on the headlands, and by 1980 it necessitated a return to ploughing. This was a great pity as the soil structure had improved greatly under the no-plough regime, and it had become much easier to achieve good seedbeds. It was helped further by the new drainage and by sub-soiling every three years.

The yields on the continuous wheat field were not spectacular, and as expected were on the low side of 30–35 cwt per acre for the first three years. Thereafter they rose steadily, especially after the introduction of fungicides for eyespot control. Towards the end, they were averaging 52 cwt per acre (6.5 tonnes per hectare) for the whole run, with a peak yield of 65 cwt. These were achieved with moderate input variable costs as I was never a believer in very high inputs of fertilisers and excessive prophylactic spraying. It proved to *my* satisfaction anyhow that continuous wheat growing *was* possible, though it was by no means a foolproof or easy system.

Continuous barley growing on lighter soils is more difficult in the case of the spring-sown crop. That, at least, was our experience on the farm. The reason is probably that the crop is relatively shallow rooting, and because it is only in the ground for some five months it just does not produce enough organic matter in its root systems to maintain soil organic matter levels at a safe figure. The light soil also means more aeration and more rapid oxidation of what organic matter is there in the first place. We found it difficult to achieve respectable yields on that field, and abandoned it after a few years.

Now that straw burning has been banned, it is going to be extremely interesting to see what happens to soil organic matter levels on those fields where straw is chopped and ploughed back. My guess would be that they will not increase as much as many people might expect, and that the more that is put back in, the quicker it will break down, always providing that the soil conditions are sufficiently benign to allow micro-organisms to work efficiently.

As a footnote to this fascinating subject, my little patch of heavy land vegetable garden has not seen a spade for twelve years, and the top soil is beautifully friable and teeming with earthworms – and slugs!

These trials of new methods that we were experimenting with on our own farm were typical of what was happening in the world outside. In the earlier part of the decade some farmers undoubtedly went over the top in the cereal growing areas in their pursuit of large-scale mechanisation and ultimate simplicity of system. Cereal prices were good compared to those obtainable for livestock products, with the exception of dairying. The temptation to rip out hedges and plough every acre on the farm became too great to resist, and we still have to live with the legacy. Perhaps Mark Antony was right when he said, 'The evil that men do lives after them, the good is oft interred with their bones.' We are certainly living still with charges of damaging the environment, as if the excesses of the 1960s are still being perpetrated today, which, of course, they are not. Accusations of 'raping' the countryside or 'poisoning' the soil with fertilisers and pesticides, are still widely made. If soils *had* been poisoned from the 1960s onwards, it is strange that crop yields have continued to improve over the intervening 30 years. If such statements *were* true, the land should now be sterile and farmers bankrupt, which manifestly is not the case. The 'good interred with their bones' is the fact that cereal farmers are today producing the nation's requirement at a very competitive price, and contributing in no small way to Exchequer funds through the export market of surplus wheat and barley. At the same time British farmers have probably become the most conservation-minded in Europe.

I have tended to concentrate on developments in cereals because that was the area which caught the attention of the public more than any other in the '60s, and the area which attracted the highest proportion of the subsidy money devoted to agriculture. But, of course, it was a very similar picture in other sectors of arable farming. With both potatoes and sugar beet, there was an increasing use of larger machines and of chemical aids to counter pests and diseases and improve the quality of the product. Potato yields increased from an average of 8.5 tons to 11 tons per acre, and outdoor storage was largely replaced by environment-controlled indoor stores. Inevitably, as in other sectors, small producers began to disappear, as did beet producers on the more marginal farms or those distant from the factories. Production became more and more concentrated on large farms in those areas especially suitable for the crop, which could afford the increasingly high cost of mechanisation. Perhaps an even more significant change was taking place in the production of vegetables, where small-scale market gardeners began to find that it was impossible to compete with larger growers, and with crops grown on a field-scale with an ever increasing use of precision mechanisation. Small-scale vegetable growing was always highly labour intensive, and rising wages, if nothing else, were bound, sooner or later, to squeeze out the small man. They have not all gone, of course, for some manage to survive catering for local, or niche markets, and by working extremely hard, but thousands have disappeared from the agricultural statistics.

The other factor which helped to put a nail in the coffin of the small grower was the advent of deep freezing, which began to transform this sector in the '60s. It also enabled arable farmers to introduce another rotational crop − such as peas − into their farming systems, and exposed them to the more rigorous disciplines of contract farming for processors or multiple retailers.

Altogether, then, in both the arable and livestock sectors, those ten years from 1960−70, probably saw more rapid change in farming than any other previous decade in the history of agriculture. It was a truly exciting and stimulating period in which to have lived and farmed – at least, for those who were able to meet the challenges, and come to terms with the pace of change. For those who, sadly, could not do so for one reason or another, it must have been a time of worry and sorrow, but at least there was then a reasonable level of employment outside, which is not the case today for those who are being forced to leave the industry.

EDUCATIONAL DEVELOPMENTS

Reference has already been made to the National Certificate Exams Board and in the autumn of 1962, its Chief Assessor, Charles Mercer, who had guided it through its early years, died. Soon afterwards I got a call from Ted Stearn at Brooksby, who was on the Board, to say that they would like me to take on the job, and that I would shortly be hearing from Olive Foss at City and Guilds with an official invitation. He said that they were looking for someone from the academic world who was independent and not directly connected with the Farm Institutes and who was sympathetic to education at craft level. Though flattered to be invited, I was hesitant to say I would do it as I was dubious about taking on any more commitments. I was deeply involved with the Oxford Conference, then at its peak of popularity, and I had the YFC Council and examining jobs at other universities. Ted twisted my arm quite hard, saying that he was certain that I would have the confidence of the Principals of the Institutes – clearly an important consideration – and I already knew quite a large number of them through membership of the Agricultural Education Association. I fear that I must be like the girl in the song from *Oklahoma!* – 'I'm just a gal who can't say no'; so I said that I'd take it on, if they really did want me to do it. So began another activity that was to become a major part of my life for the next 25 years. If I could have foreseen that it was to be for so long, I might have turned it down, though I am glad that I did not do so, for it turned out to be an intensely interesting and rewarding experience.

Since I had ceased to be an assessor in 1959, the NCA exam had become further established, and there were now 32 Institutes in the scheme with some 1,300 candidates and nine assessors, excluding the Chief Assessor. The main principle on which the Board worked, of having locally set and marked exam papers, moderated and examined by the external assessor, who also interviewed each candidate at the end of the exam, had proved very successful. Already some of the Institutes that ran other courses were asking the Board for a similar type of examination to give them national recognition. So these were urgent matters to be dealt with, quite apart from establishing contact with the assessors who would be working under me. The first task was to put in place a new certificate for girls. There was already a considerable number in some colleges, who were doing home economics, and who had no intention of trying to get full-time jobs on farms. Many of them were farmers' daughters, possibly later to become wives of farmers. They really wanted training in animal husbandry (as they might do part-time work rearing calves, looking after poultry, etc.), in book-keeping and accounts and farm records, and in domestic science, which by then had been renamed home economics. A pilot scheme had already been run at three colleges that year, and it needed tidying up and formalising before the Board could launch it as an official examination. It meant visiting a few Institutes to gain first-hand knowledge, but there were few problems, except the knotty one of how much farm machinery should be included. As the exam was structured to include major and minor papers, that was fairly easy, since machinery could be demoted to a minor subject, and Institutes left to teach the sort of machinery elements that might be useful to girls in that area. Girls who were keen to get full-time jobs on farms still took the ordinary one year farming course.

It all went quite smoothly, and we were able to put in place a Certificate in Agriculture With Home Economics the following year. It meant finding lady home economics assessors to work in partnership with the male assessor in each case. We started with three assessors, and eight colleges with about 100 candidates, and over the years the numbers remained constant until horses came prancing onto the scene in the late '70s, and took some of the girls away. Apropos of the home economics exam, I came to learn quite a lot about the finer points of cookery, the removal of stains, the washing of undies and similar topics. This was through having to read the draft examination papers each year and make any suggestions that I thought necessary for clarification in cases of sloppy wording. Often there was quite a lot of re-writing to do, as I could not possibly let through a question like 'What steps can a dairymaid take to improve her milk yields?' which appeared on one occasion. Actually, it would have been interesting to have left it in, to see what the answers were!

The other major task I was given was more complex. This was to report on the feasibility of devising a scheme to cater for many second year courses which had sprung up in the Institutes over the previous ten years, as the demand for advanced specialised education expanded. The reason for this was the growth in size of individual farm enterprises and the need that farmers had for well-trained workers to take on the responsibility for running them on a day to day basis. There was also a growing need for training in full farm management both for farmers' sons and for others with management potential, who for one reason or another could not attend courses at the larger colleges. 'Management', as I have said before, had quite recently become the in-word in farming circles.

I found that there were some 20 colleges offering a wide variety of advanced courses, and when I had looked at their syllabuses and visited quite a few of them, it became clear how different some of the courses were from one another. Some were only of two terms' duration and extremely specialised, such as a course in pig husbandry, whereas others covered general husbandry at a more advanced level than the one year course. Others again were almost entirely devoted to farm management. I decided that it would just about be possible to formulate valid examination schemes for two main groups – those courses concerned with husbandry, and those concerned essentially with management. So I produced two blueprints for the Board, and worked out detailed examination structures for each. These were based on the same principles of two major papers, three or four minor papers, and a third section covering practical assessments by the college staffs throughout the course – a system that we already had in the main NCA examination. The certificates to be awarded would be in advanced husbandry in the first case, and farm management in the second.

One thing that I was most insistent about was that in the minor paper section there should be what I called a project study. Every candidate would have to select a topic in which he or she was interested, investigate it in depth and possibly have a practical farm experiment on it, and write a dissertation of some 3,000 words, to be handed in well before the date of the examination. This arose from my experience of how valuable I had found my own Cambridge project to have been. The preparation of such a report is an excellent discipline in learning where and how to look for information, how to assemble facts and relate them to one another and finally how to get them down on paper in a clear and orderly way. This can be a much better assessment of a student's basic ability than storing up a lot of facts from lecture notes and then spewing them all back again in a set exam paper. Projects are now, of course, part and parcel of many examinations, both at school and college level, but 30 years ago, they were not at all common. For the farm management certificate I decided that a full management study should be required for one of the major papers in the final examination.

The schemes proved to be acceptable to the Board and, in general, with the Institute Principals who were involved. We had a conference with them and ironed out a number of relatively minor issues which they were concerned about, and the first examinations were held in 1965. I decided to assess about half the husbandry courses myself and asked the late Ray Mortimer of Harper Adams to do the others. Ray was a first-class teacher, and a true all-rounder, and we worked very well together. It was a real tragedy and a loss to agriculture when he died in his early fifties some years later, from cancer. So within two years, instead of just having one examination to look after, I had four, each with specialist assessors, and it was already becoming a larger commitment than Ted had led me to expect.

The important thing was to establish a speedy foolproof system of dealing with all the question papers that came flooding in for approval in March/April, and then for handling the already marked papers from the colleges at examination time. It was essential to build up good personal relations with our customers, the Institutes, and to make them feel that we were an efficient organisation, which didn't waste a lot of time on bureaucratic red tape. We were extremely lucky to have Olive Foss nominally Secretary of the Board at City and Guilds, and between us we established an effective routine. She provided me with a succession of very efficient young assistants who were responsible for the day to day running of the schemes at the secretarial level, and I do owe them a tremendous amount for their unfailing goodwill, co-operation and friendship over the years, especially in those early stages of expansion.

And it *was* a period of expansion, too, for within a few years two Scottish colleges and three from Northern Ireland had all joined the National Certificate Scheme. By 1970, I had 40 Colleges and Institutes participating. We had an excellent forward-looking Board under the chairmanship firstly of Henry Mason, of the NFU, and then later of Peter Fox, a Yorkshire farmer and Treasurer of the NFU. They were prepared to let me get on with running the scheme as informally as possible. If there was a problem, I would ring the Chairman, and it was his decision whether it should wait for a meeting of the Board or not. Between us we were often able to sort things out and give the college concerned a quick answer, which was certainly appreciated. Over the next 15 years, we established new examinations for farm secretaries, and then for the management of horses, as the number of courses concerned with horse husbandry and management proliferated at the colleges in the later '70s and early '80s. These were all based on exactly the same model as the original examination.

From a personal point of view, there were two pressure

periods, the first one in March/April when the proposed examination papers submitted by the Institutes were approved, and then in late June/July when the examinations were held and the candidates had to be oralled. This suited me well, as by this time I had always put the year's Oxford Conference to bed, with the report at the printers, and our students were away for the Easter vacation when the papers came in. The later period came just after the Oxford term had ended, and there was some time to spare.

From the start I made a rod for my own back by insisting on reading through every question paper myself, even though the assessors would do this for their respective colleges, and send me any comments that they might have on them. It wasn't that I didn't trust them to pick up obvious errors of overlapping questions, sloppy writing and ambiguities, but two heads are better than one, and it was surprising how easy it was to miss important points. It was rather that, as we were running a national examination, I was determined to set the highest standards for grammar, and clarity of objective in the setting of the questions. I probably got something of a reputation for 'nit-picking' in some of the colleges when I insisted on decent grammatical English. But in an era of increasing laxity in such matters in education, I felt that a National Board should adhere to good levels of English. Some colleges welcomed this as a sign that the Board was doing its job properly, in ensuring comparability of question setting right across the spectrum of 40 colleges, which was clearly an important consideration. But by the time I came to hand over, the job of reading through 87 sets of papers, with usually six papers in each set, was really getting to be too much of a good thing.

The other pressure period was when the marked scripts arrived, usually only a couple of days before the orals were due at the college. We tried to save travelling expenses by doing two or three colleges on a round trip. There was never time to read every answer. One was only checking on the validity of the internal marking to see that it was of an adequate standard, and one that was comparable with the marking at the other colleges on one's list. But I always told assessors they should read the papers thoroughly for those candidates who were on the borderline for the three grades in the examination, so as to form an opinion about such students before meeting them.

The popularity of the old NCA examination (and I have to say 'old', since I gather it has now changed completely) was due to two things. Firstly, the quality, patience and dedication of the team of assessors, and their willingness to co-operate positively with the staff at the colleges, especially when it came to the grading of their students. They really did do outstanding work on behalf of the Board. Secondly, to its informality and absence of bureaucracy. If a college had a problem, they knew they could ring me or the secretary at City and Guilds, and they would get an answer on the phone if we could give them one, and they knew where they were. Only the other day, a college principal said to me, 'I look back with nostalgia to those days when we could ring up and get an answer, whereas now we have to write endless letters and are lucky to get a decision within six months.' In retrospect, I suppose that I was a bit of a dictator in handling so much of the routine business myself with the secretariat. But the Board trusted me to keep the show running smoothly, and the colleges and assessors welcomed the personal contact rather than dealing with a remote official. For my part, I derived great pleasure from the friendships made in the colleges and from the warm welcome that I always received from the staff. They were good days.

It was early in the year after I had taken on the NCA that I got a call from Olive Foss asking me to come up to lunch one day as she had something to discuss with me. It transpired that City and Guilds had decided to expand its activities in the agricultural and horticultural sectors, and to unite all the different interests, including machinery courses (which had previously had a small committee of its own) under one major Advisory Committee. This would bring agriculture and horticulture into line with all its other major industrial advisory committees. They were anxious that I should chair it – again as someone outside the regular college field, but with a first-hand knowledge of farming. As was the case the year before, I was very hesitant about taking it on because I did not know a great deal about the part-time courses, or many of the people involved. A lot were run from technical colleges and not Farm Institutes, even though the local principal was often nominally responsible for them. Olive insisted that they wanted me to do it to form a close link with the full-time courses and the NCA Board. She assured me that she had made enquiries round the centres and also the Department of Education and that I would be very acceptable to them as an independent 'outsider'. She also said that as soon as it was established her staff would do most of the work, and it would really only mean chairing about four meetings a year and attending a few sub-committees from time to time.

Olive is a persuasive lady, and once again I fell for her blandishments, weakly saying that I'd take it on, if nobody else was proposed at the meeting. Of course, nobody was, as Olive had engineered it all beforehand and persuaded two influential members to propose me, and I suppose that nobody dared to oppose them by putting another name into the ring! It turned out to be a big committee of well over 30 members, mostly representatives from every organisation remotely connected with agricultural and horticultural education. They included people from the trade unions, teachers' bodies, the RASE, the NFU, Local Authority Parks and Gardens, representatives from Wales, two HMIs, and a great many more besides.

These big representative committees are not very easy to manage since the interests of the different bodies represented are so diverse. Very often, a matter under discussion is of no importance to half of the people there, and they tend to get the feeling that they are just there to rubber stamp decisions which do not concern them. Accordingly they can get rather easily bored, especially as with a committee of that size there are not the opportunities for many people to speak, because of the numbers present. I don't know what the answer is, except to keep committees as small as possible, but then there is a danger of some organisation feeling excluded when a matter that might concern it comes up. The other thing is to try to see that there are diverse items on the agenda, which will keep quite a number of the members involved.

It immediately became obvious that Olive's forecast that the appointment would not entail very much work was very wide of the mark. Even at the first meeting it was decided that the time had come to pull together all the syllabuses which had grown up piecemeal over the previous ten years, and to produce new ones across the board in a standard format. Initially, these were to cover first year animal husbandry, crop husbandry, and machinery operation and care. The horticulturists would have their own sub-committee to produce their syllabuses. When these were finished, the next step would be to produce a set of more advanced syllabuses to cater for specialist second year courses. It was also decided that all of them should give as much emphasis on practical work as possible, bearing in mind the shortage of facilities for practical instruction at many of the teaching centres.

It was clear that small specialist working parties would be needed for each subject but there would have to be a strong co-ordinating force overseeing them. This would ensure that they were written at an equivalent level, were in a similar format, and had roughly the same balance of theoretical to practical instruction. Since the staff at City and Guilds, competent as they were, had little if no knowledge of agriculture, they could not be expected to fulfil that role, which meant that the only person who could do it with authority was the chairman. So I found myself chairing three working parties, which turned out to be a mentally stimulating experience, and quite a challenge, too.

An effective working party ought not to exceed four people, and that may be one too many. I was anxious to see that ours should be composed mainly of youngish people actually involved in day to day teaching at ground level. I wanted to avoid senior people because they might not be able to find the time to prepare drafts, and because the more senior people become, the more likely they are to be caught up in administration and be out of touch with basics.

We had previously had a lengthy discussion in the main committee whether we should have a first year course in general agriculture, and it was decided not to do so, as with only some 32 weeks duration at one day a week, it would have been so superficial as to be practically worthless. So we went for the three subject approach of animals, crops and machinery, and set up the three working groups accordingly. They did really sterling work, with personalities such as Tony Harris, Bob Nelson and the late John Turner producing the drafts for us to knock into shape. I suppose that it took us nearly a year to produce the comprehensive scheme for stage 1, and then it was time to tackle eight more specialist subjects at stage 2 level. When that was finished we prepared a stage 3 for elementary management.

The syllabuses were generally well received, though some teaching centres still preferred to stick with their own courses under some of the Regional Examining Bodies. That was fair enough at this level, where perhaps a more widely accepted national qualification, such as a City and Guilds certificate, was not so important as at the full-time course level. Throughout all the discussions that took place at the time, and for many years afterwards, two topics kept on recurring. Firstly, the difference between education and training, and secondly, the role of proficiency tests, and what was meant by proficiency in the first place.

There is no hard and fast line between education and training since training is, after all, one form of education, but in constructing a syllabus at craft level, it is vital to try to clarify the terms, and lay down guidelines for teachers. I would always try to rationalise it when the arguments got too heated in committee, by saying that education is basically informing a student *why* a particular task should be done, i.e. providing background knowledge, whereas training is teaching the student *how* it is done. This in its simplest form means the difference between theoretical knowledge and practical performance. In the first syllabuses that we produced, we had to include a good deal of theoretical material, because the facilities for practical instruction at many centres were still very inadequate. At the next major revision some eight years later, things had improved very considerably in that respect and we included much more practical instruction and lists of tests to be performed.

That leads on to proficiency. In the 1950s the National Federation of Young Farmers Clubs, had pulled together a number of practical tests which were being used in some of the clubs, in a standard format. They were concerned primarily with machinery operation and livestock tasks and it was important to set criteria so that examiners in different areas would be working to the same standards. It was not long before some teaching centres sought to incorporate the tests in their courses, and the whole concept began to expand. It was clear that the NFYFC was not the appropriate body to run a large scheme of this nature, and anyhow, they had not the staff to do it. So in 1970, the National Proficiency Tests Council was set up, composed of representatives of the teaching organisations, the unions, the agricultural and horticultural bodies, and so on. This became the body for setting standards, devising new tests, co-ordinating examiners, training new examiners and setting fee levels. Its secretary, Philip Sheppy, very ably guided its development and progress for 24 years, in spite of many difficulties, till he retired in 1994.

The problem as it concerned my committee was whether we should include one or more of their official tests as a mandatory component of our more advanced examinations. The difficulty lay in the different interpretations that people put on the word 'proficiency' as applied to the performance of jobs on the farm. In acquiring manual skills, there are basically three different stages: the first is simply acquiring the ability to perform a task to an elementary standard when under instruction, then there is the ability to perform that task to a reasonable standard when not under supervision, and finally, there is the ability to do it routinely to a high standard day in, day out. The first stage can be called 'ability to perform', the second can be labelled 'competence', and the final one, 'proficiency' – the standard expected of a qualified craftsman. But to reach the final stage takes a considerable amount of time and practice – far longer than that available on a 120 hour, 30 week course. So the tests could really only apply to those candidates who had the opportunity to get a lot of practice on the farms on which they worked and would exclude those who could not. There was a further risk that if we did make them a compulsory element, the results obtained would not be available by the time we wanted to issue our certificates. So, we did not include them in our initial syllabuses.

Another aspect of the job was to chair the so-called Moderating Committees, which met each autumn to approve the question papers to be set the following year. We dealt with four or five papers at one meeting, and I would invite to it one or two members of the full Committee and the examiners who had prepared the drafts of the papers under discussion. I was warned beforehand that moderating papers in the machinery subjects always took a very long time and gave rise to heated discussions. So a firm hand was needed, as I was not prepared to miss my train home, because of excessive argument. It was surprising how quickly the firm hand worked, once I had made it clear that I expected reasonable give and take, and that if they couldn't agree, I would give the ruling, which from experience I was generally in a position to do with reasonable authority. The staff of City and Guilds are, of course, highly trained in the theory and technique of setting questions, and I learned a great deal from them which was very useful when it came to moderating all my NCA papers each year.

One cannot leave this period without a mention of the battles which took place following the passing of the Industrial Training Act of 1964, which led to the setting up of the Agricultural Training Board. Industrial boards were established to organise training courses, often in association with the Local Education Authorities, to take over apprenticeship schemes, to set up training groups, and so on. The key point was that they were to be funded by levies on the employers. In the case of very big industries, it meant that the larger employers had to pay the bulk of the levies, which was reasonably fair if they were not already running training schemes of their own – as many of the best employers were. So it was not popular with them, but the fuss that they kicked up was as nothing to that created by farmers, when the time came to collect levies from them. Smaller farmers with only one or two men did not see why they should pay for other larger farmer's employees to be trained, and many farmers took the view that their men were already very skilled in their jobs and didn't need training. I don't think that I ever remember an issue which had the farming community so much up in arms, with people who refused to pay being taken to court, and big demonstrations by normally law-abiding farmers across the country. Eventually, of course, it was settled by arranging that the Board should be financed out of the funds made available at the February Price Review, which took it out of the public eye, at any rate.

A local Area Training Committee was set up to cover Berkshire, Buckinghamshire and Oxon, which I was invited to chair. It was a difficult committee in the prevailing climate of hostility, and I was glad to give it up after a couple of years, when they seemed to want a commercial farmer as chairman. The most successful of the ATB's activities, during the '70s at any rate, was the formation of small local training groups. These were composed of enthusiastic farmers who could see the value of keeping both themselves and their employees up to date with both managerial and technological change – and that still applies today.

I seem to have spent a good deal of time on the theme of education, but it did monopolise quite a lot of my extra-mural activities in the '60s – and in the next decade, too. It was a great experience to have been so closely involved with it in different ways during the period when it assumed such importance in the technological revolution that was taking place in farming practice. A very rewarding involvement it was, not least because of the outstanding quality of many of those in the agricultural education service. I received much help, not only from the teachers and principals, but also the members of the staff at City and Guilds, who, working under considerable pressure, never failed to come up with new drafts at short notice.

The Institute at one time in the following decade became almost a London club for me, as I seemed to be lunching there about once a week. I had by then been put on the Institute's Senior Awards Committee, which meant another four meetings a year, dealing mainly with its Insignia Award covering all the industries that City and Guilds catered for.

My old friend Peter Stevens, who was one of Olive's first assistants to be deputed to look after agriculture, and who 25 years later became Secretary to the Institute, told me that the staff always enjoyed looking after agriculturists because they were so pleasant and easy to work with. He said that they brought a wave of fresh air in with them (one hopes that it was not tainted with farmyard manure or silage effluent!). He was right, for agricultural education for the past 40 years has been blessed with an exception-ally dedicated and committed group of men and women who have always put the welfare and interests of their students first. They have gone to great pains to bring on and encourage the backward ones, and to stimulate the more gifted ones. There has never been a nine to five attitude in agricultural education, and I hope that there never will be. I know this to be true from the insight that I gained from meeting college staffs over 25 years of acting as an NCA Board Assessor. They were always able to give detailed information about their students' backgrounds and problems, if it was relevant to the performance of a candi-date. I do count myself very fortunate to have had the opportunity to be so closely associated with them.

A CRISIS IN THE DEPARTMENT

It is time now to return to the situation in our own Department at Oxford, which had begun to look rather precarious in the early '60s due to the small number of students applying for the course. The University Grants Committee (UGC) quinquennial review and visitation was due in 1965, which was an added cause for concern. The Forestry School, too, had a small intake because the Empire was no more, and developing countries had begun to educate their own nationals. Forestry did have a successful one year postgraduate course, but we only had one official one in soil science, which was not heavily subscribed. We did have a very good first degree course. Most colleges in the University were happy to accept undergraduates for it if they were academically of an equivalent standard to those admitted to read zoology or botany. However, the schools were just not entering candidates because the word seemed to have got round that the standard required to get into Oxford was extremely high, and that entrants for agricultural science did not stand much chance of getting a place.

It was about that time that school parties started visiting the Oxford science departments, and we went out of our way to put on interesting programmes for them. One of the members of staff – David Smith, later to come back as Professor – wrote a very useful article for the *Times Educational Supplement*, describing the course. Whether these two things helped, I am not sure, but by the time of the UGC visitation in 1965, the numbers had picked up a little, though they were still dangerously low.

Rumours had been flying around for sometime beforehand that the UGC Committee intended to close three university departments of agriculture. The suggestion was that there were too many in relation to the likely future requirement for graduates and that money would be better spent on concentrating resources on fewer of them. We, as the one with the smallest numbers, were generally thought to be for the chop, and Cambridge, in spite of its pre-war reputation, was reputed to be another. Although their numbers were larger than ours, the course had changed little over the years, and was felt not to be of a high enough scientific standard. Someone even came up to me one day and said, 'What are you going to do, Mike, when your department is closed down?' My reply, as I recall it, was that it would be time enough to think about it if the axe did fall. But I didn't see why it should, since it was a good course in line with modern thinking about scientific agriculture, with a good practical backing.

During the visitation, the Agriculture Sub-Committee of the UGC, which had about eight members, first interviewed Geoffrey Blackman (having already been provided with a mass of background information), then members of the staff together, and then representatives of the undergraduates and postgraduates. We took good care to see that the latter were a sensible cross section of the student body, who were in tune with the ethos of the Department. At the meeting with the staff, the Committee members were all lined up on one side of a long table, with us on the other side. The Chairman of the Committee was none other than my old tutor and mentor, Jem Sanders, who was by now Sir Harold Sanders, the Vice Chairman of the full Grants Committee. I also knew two of the farmer members well. One was an ex-chairman of the Farming Conference Committee, and the other a member of the same committee. In addition there were two professors whom I knew well as fellow members of societies to which I belonged. In fact, there were only two members that I did not already know well. I would not suggest that this in any way influenced the issue, but it greatly helped to break the ice, and get the meeting off to a friendly and informal start. They had a lot of questions to ask us about aspects of the course, the need for the type of graduate that we were producing, the attitude of the colleges, and where we thought that the course might go in the future. We were a good team at the time, who knew each other very well, and we managed to throw the ball about a lot among ourselves in discussing their questions. I was told afterwards by one of the Committee members that this made a very considerable impression on them and that we came across as an effective and united team with a positive and progressive outlook — and accordingly we survived! In the subsequent report, it was recommended that we should make every effort to get more students, and that the university should consider integrating the agriculture and forestry components more fully. This was achieved four years later when the course was reorganised and became a degree in Agriculture and Forest Science. By then the numbers had begun to increase rapidly, with the course becoming more popular in the schools and our own colleges.

The rumours regarding Cambridge proved to be only too accurate, and the support for the agriculture course there was withdrawn, and the university converted it into an applied biology degree. Naturally I was very relieved that we had survived but was much saddened by the demise of my old university department. Only 30 years before it had been in the forefront of high level agricultural education and the workplace of so many of the leaders in scientific research between the wars. But if it was a sad day for me, how much more so must it have been for Jem Sanders, who, as Chairman of the Committee, had virtually to sign the death warrant for his former department in which he had earlier done such outstanding work. Years later, when I visited him shortly before his death in 1989 I asked him about this, and he admitted that it was one of the most unpleasant jobs that he had ever had to do, but that there was really no alternative at the time.

Geoffrey Blackman came up for retirement in 1970, and I had never had any regrets about going to work for him in 1946. He was a delightful person to be under, and one who let you get on without interference so long as he felt that you were doing your job. He was never an undergraduate's professor, being first and foremost a distinguished scientist. He was more at home perhaps within the portals of the Royal Society or the Athenaeum than on the farm, and he was intellectually too far above most students for them to be able to relate easily in his lectures. Though he spoke quite widely at farmers' meetings in the earlier days of herbicidal expansion, here again, he was inclined to pitch his level rather above the heads of the average farmer, though the very progressive ones held him in high regard. He did excellent work on herbicides and their application on farms in the late '40s, and he was very far-sighted in his support for oilseed rape and forage maize. I hope that he can now look down from some far corner of the Elysian Fields and view with pride the fantastic expansion which has taken place in the acreage of both crops in recent years.

As Geoffrey was away a great deal it was important that there should be someone in the Department of relatively senior status to whom undergraduates could relate, and who could act as a kind of staff officer. I fairly quickly came to take on that role, allowing him to concentrate on research, university politics (very important, this), and work on scientific committees, which is also important from the point of view of raising money. In the modern world holders of Chairs have so many responsibilities in such fields and are so concerned with raising funds to ensure the survival of their departments that they seldom have much time to devote to undergraduate affairs. It is sad, but inevitable. While on the subject of Chairs, I was head-hunted a couple of times during the '60s in regard to accepting one, and in one case put under a good deal of pressure, but I was aware of my limitations for such a post and said 'No, thanks very much'. Firstly, because I was never a very good scientist and thought that I might find it difficult to raise adequate funds for postgraduate research, and secondly, because I knew very well that if I accepted, I would become a desk-bound administrator, and might just as well throw away my wellingtons, as I would not be able to get out onto a farm when I wanted to. There was also a third reason – that I was never one who was hungry for power, and was certainly not persuaded that there was a great deal of merit in being called 'Professor'. I had taken on the NCA and City and Guilds work fairly recently, had the Oxford Conference to run and was a member of several other committees, all of which would have to be given up if I accepted the offer. So why on earth abandon all these things which I enjoyed and have the hassle of starting a new life elsewhere? Certainly I never came to regret that particular decision.

I have already mentioned several times the specialist societies which had sprung up at the end of the war. Of the three to which I belonged, it was the meetings of the British Society of Animal Production, founded in 1944, that I most enjoyed, and which were the most relevant to my teaching. The summer conferences were especially valuable as they moved around the four countries in the UK, giving one an insight into farming conditions in all parts of the British Isles. The Society had a very strong Scottish element, having been founded largely in Edinburgh, and it was very fortunate to have had two excellent Secretaries in the early days. These were John Maule and the late James Walker Love, who established a pleasant atmosphere of friendliness and informality, which made the conferences most enjoyable occasions, both the peripatetic summer ones and the spring paper reading sessions then held in London.

In 1960, someone suggested that I should stand for a place on the Council, to which I was duly elected, and then suddenly found myself two years later nominated for the Presidency. This was a tremendous honour, following in the footsteps of distinguished scientists and administrators. The summer conference in 1963 was due to be held somewhere in the South, so we suggested that the members might like to come to Oxford. I got together a strong local committee to plan it, and managed to book Christ Church to house us. It all went off very well, with a lot of interesting farm visits, and a good attendance. One thing that I do very clearly recollect was having to make a dinner speech from High Table in the Hall, which is by far the largest, and most impressive of all Oxford college dining halls. It is so high that one's voice just disappears into the beams, though strangely enough, it is far better when speaking at ground level, below High Table, as I was to find on another later occasion.

The Society has, of course, grown astronomically since then, and acquired an international reputation. I well remember the discussions that we had when I was on the Council as to whether we could run the financial risk of starting an official journal, which, in fact, we did, and it was very well received. Today, I get six instalments a year, each one of about 150 pages. Never in our wildest dreams could we have anticipated it taking off in such a fashion.

So the decade of the '60s closed, and a busy and exciting time it had been, both for farming, and for me personally. Unpredictable times were looming up ahead, with a possible entry to the European Community already being discussed, with all the implications regarding agricultural support that were bound to come with it. In addition, thinking people were already becoming concerned once again at the levels of production being achieved, and fearing that some kind of restriction would become necessary. It was to prove to have been an era to which those who farmed through it will look back nostalgically as a period in which they were able to farm as they liked and to expand as much as they wished. It was a time when inflation was something which was of little financial concern, and bureaucracy was minimal compared to the regimented world of today.

CHAPTER SIX

The Decades Pass – 1970–80

───※───

HAVING survived the threat of closure in 1965, the Department gained a new lease of life with a rapid increase in student numbers, helped possibly by the closure of the Cambridge School. By the time that Geoffrey Blackman came to retire in 1970, the course was lively and active once more, which made teaching much more fun again. The recommendation from the Grants Committee about integrating with Forestry was implemented in 1970, and we then had a course which was unique in the country, combining agriculture and forest science. It proved to be particularly valuable for those who hoped to work overseas in developing countries (and there were quite a number of them). It also provided a wide coverage of plant sciences, with a number of options which suited the needs of the range of students that were attracted to it.

The question that concerned us all at the time was who we were going to get as the replacement for Geoffrey. The hot favourite was John Burnett, a former Oxford botany graduate, who was a well-known mycologist who had been a Fellow of Magdalen College for some years while working in the Botany School. He had then moved on to no fewer than three Chairs of botany at Newcastle, St. Andrews and Glasgow. The tipsters were proved to be correct, and he was duly elected to the Chair early in 1970. A Scotsman by birth, who had been educated in England, John Burnett was a very different type of personality to his predecessor. He was much more representative of the modern administrator breed of professor, rather than the older type of scholarly academic.

He had quite a good knowledge of agriculture — certainly enough to get by as a professor of agricultural science — and in two of his earlier posts he had worked in universities with strong departments of agriculture. In addition as a mycologist he had worked with plant diseases. Another strong advantage was that he knew the Oxford scene very well, which is a great help when it comes to staking claims on behalf of a department in university affairs. Personally he turned out to be very approachable, and very quickly established his position in the Department, and was more student orientated than Geoffrey had been in his later years.

Naturally I was very concerned as to what attitude he was going to take with regard both to my numerous external activities, and also to the freedom that Geoffrey had allowed me in running the farm very much as I liked. So, soon after he arrived, I knocked on his door and asked if he had a moment to spare. I dived straight in at the deep end by saying, 'I'm not sure if you know that I do have rather a lot of commitments outside the university in the educational world, and sit on a number of agricultural committees, apart from running the Conference each year, so that I do seem to spend quite a bit of time in London.' He laughed and said, 'Well, I was, of course, aware that you do a lot of outside work in various ways. But I've been going through the timetable, and see that you seem to do nearly twice as much undergraduate teaching as most other members of staff. If you can fit that in with your committee work and the farm, that's fine by me, as it all helps to keep the Department on the map.' So that was a relief, as I didn't want to give anything up if I could help it. John proved very easy to work with, so once again, I was fortunate in my boss, as I have been throughout my whole career — and what a difference that must make to contentment in life!

He initiated a number of changes in the Department, one of which was the introduction of compulsory field courses for the students, two of which they had to attend over two years on the course. Geoff Hodgson and I were very enthusiastic about this, recalling the good times we had on the Scandinavian trips when he was still an undergraduate. We ran our first one in October 1971 at the Malham Field Studies Centre in the Yorkshire Dales, just before the start of term. But Geoff thought that we ran a risk of bad weather up there as late as that in the year, so thereafter we went to Dartmoor instead.

With the help of that great personality Ian Mercer, then the Chief Executive of the Dartmoor National Park Committee, we found a privately run Field Centre near Bovey Tracy which suited us very well. It was reasonably inexpensive, and well placed for the range of topics that we wanted to deal with. These covered the ecology of upland areas, land use and the competing claims for land in a National Park, upland farming, forestry, tourism, mineral extraction, and a day's visit to lowland farms. Other members of staff joined us for a day or two for special subjects, and I built up a group of outstanding farmers to visit, originally with help from the staff at Seale Hayne. We worked hard, with inquests after dinner each day on what we had seen, or an occasional outside speaker such as Ian Mercer, who was excellent value and a mine of information. Geoff and I found them great fun to run, as we had vintage groups of students throughout the '70s, who worked together extremely well, and they appeared to enjoy the trips as much as we did. I like to think that they learned a tremendous lot in those seven days, especially from the farmers, who always seemed to be delighted to see us each year, and to whom I owe a great debt of gratitude for their hospitality.

In retrospect, the last ten years of my teaching life were just as pleasurable as the first ten had been in the post-war years, largely because of the quality of undergraduate attracted to the course, and the number of natural leaders that materialised each year. There was one marked difference between the two generations apart from age, and that was in their gender. In the post-war years, we were fortunate if we managed to get one girl a year on the course, whereas in the '70s the proportion of girls steadily increased until it came to about one in three, which was very civilising for all of us!

After about three years, we saw very little of our, by then, not-so-new professor, as he had thrown himself into university affairs with such abandon that he soon found himself elected to the office of Vice Chairman of the General Board (effectively the Chairman). The General Board is the administrative committee which runs the day to day business of the University. He got leave of absence to do that for three years, and quite soon afterwards was appointed Principal and Vice Chancellor of Edinburgh University. So, by the end of the decade, we were back where we started, awaiting the appointment of another new professor; but by that time I was not very concerned as I was due to retire in 1981 anyhow.

CHANGES ON THE FARM

In the farming world outside, things were still changing rapidly, as indeed they were on our own farm at Wytham. As before, the sort of problems that we were encountering there were very typical of those confronting many other farmers. In our case, the major event was the death of our neighbour in 1970 which released quite a large amount of land. The university land agent very sensibly suggested that we should take on the extra 220 acres, which meant that we would then be farming some 640 acres in total. Our neighbour's death was really a tragedy, as he was still a relatively young man, and it left us very saddened. To me, it exemplified in an extreme form a scenario that I had come across several times during the war when dealing with difficult cases on behalf of the Committee. Our neighbour had lived on the farm all his life, working under a dominant father from the time that he left school and had had no formal training. I suspect he had never been given any responsibility either so when his father died suddenly he was left floundering. I think the university had made a mistake in saddling him with another 100 acres when our other neighbour had retired six years earlier. It was just too much, and he worked himself to death trying to keep on top of it. The saddest thing was that one could see what was likely to happen, but could do nothing about it, except to lend him labour from time to time, when he needed extra help. Thank goodness for today's widespread opportunities for agricultural education, and for the enlightened attitude of farmers who encourage their sons to make use of them.

By this time, a progressive tenant on the other half of the estate had also had two smaller farms added to his, so that seven smallish farms on the estate in 1950 had, by 1970, been whittled down to three. Some ten years later they were down to only two, as another neighbour retired, and his farm was divided between ourselves and the other tenant. This was very typical of what was occurring across the country as smaller ill-equipped farms on not very good land succumbed to economic pressures, and landlords sought to rationalise the size of their farms and to get good tenants for larger efficient-sized units.

So, in our case, in 20 years, we had doubled our acreage, and were farming it with a smaller labour force, though the capital investment required to achieve this was very considerable. In economic terms, the university was now demanding a rent for the extra acreage, but we were continually enhancing the value of its property by investing profits in new buildings. So they were batting on a good wicket, so long as we were able to maintain those profits, something which got steadily more difficult over the next decade.

Of the new land we inherited, there was 70 acres of arable on heavy land, which adjoined our existing block, and 170 acres of permanent pasture, on the far side of the village, which extended out well beyond the Oxford Western by-pass. The access to the pasture was quite difficult as it was at least a mile from the main buildings. The arable was all in need of drainage, and the pasture was either wet, or very wet, reaching down to the Thames. I was dubious about ploughing any of this because of the risk of flooding. However, it is clear from the recent management of the land that I was unduly cautious, since much of it has now been ploughed and cropped, as anyone using the A34 can see – if they dare to take their eyes of that busy road at 70 mph or more!

That type of alluvial soil is very difficult to manage under the plough, due to its crumb structure, and plant establishment can be quite tricky. It is particularly treacherous when direct reseeding, as I learned during the war on several occasions. This is because the soil is so spongy that one cannot easily get a firm enough seedbed for grass and clover seed, which may germinate satisfactorily, but then fizzles out in a dry spell, as it has not got its roots down properly. So we decided to keep much of it as pasture, to get rid of the massive infestation of buttercup and other useless weeds, and to farm it on a fairly extensive system.

Our solution to the problem of how to use the extra pasture land was to increase the size of the dairy herd, which was very profitable, scrap the barley beef unit, and to use this pasture land for rearing all our own cattle. Both heifers and steers were taken through to 27–33 months, the beef cattle going out at 550–650 kg, and the heifers, most of which were sold as down calvers, at about the same age. The heifers were then very well grown and fetched a good price in the market. Both groups were reared simply on pasture, with silage in the winter, and no concentrates after the calfhood stage. With our system of both autumn and spring calving, this fitted in quite well, though we had to build some simple yards for wintering at that end of the farm, as the existing buildings on the new farm were hopelessly obsolete. This enabled the university to let the old buildings for other purposes, as well as the farmhouse, which we did not need. Many other landlords must have done the same over the past twenty years. The only difference is that they will have sold off houses and buildings and reinvested the money in their estates. The new yards adjoined the arable land and the farmyard manure was very valuable for helping to get it back into condition again.

But that left the problem of accommodating the enlarged dairy herd, which was to expand to 110–120 cows from 80. The existing yards, originally designed for 40 cows, would clearly be too small, as they were already overstretched. It meant that we either had to extend the present straw yards, for which there did not seem to be much room, or start again on a new site. I was not very keen on extending the straw yards in any case, as we were finding it quite expensive in weekend overtime in bedding them down.

It was in 1964 that a Shropshire farmer, Howell Evans, had

been nominated for the Massey Ferguson Award for developing his concept of cow cubicles, housed in large umbrella buildings or in converted yards. That principle had already been extended to so-called kennels, where the rows of cubicle standings were covered by low roof shedding. This was obviously a lot cheaper to build than large cattle yards. As we had only limited capital available, it looked as if a complete change of system incorporating kennels on a new site, which was available beyond the self-feed silage barn, would have to be the solution. I was certainly loth to abandon the straw yards which I had been used to since the days at Cambridge. It seemed to me that the cows must be a great deal more comfortable on straw than on a hard bed, even if they did have that bed to themselves, which was claimed to be an advantage – though how anyone could prove that, I don't know!

There were already rumours in circulation that some of the early pioneers had experienced difficulty in getting some cows to use the cubicles, and that foot problems seemed to be more common than with cows in yards. But it was then said that this was probably due to the dimensions of the cubicles not being quite right for the size of the cows, or that the cubicle divisions were at fault, allowing some cows to get trapped, or making it difficult for them to lie down. The height of the bed was also blamed for some foot problems with the cow's foot slipping as she stepped up. Still another disadvantage was said to be more difficulty in spotting cows in season, as one cow may be unwilling to jump another cow on slippery slurry-covered surfaces.

In regard to the point about dimensions, I could never see how one would be able to get this right with a herd of big cows like Friesians or Holsteins, so long as one had first-calf heifers, middle-aged cows, and big heavy old matrons all using the same dimension of cubicle. I was certainly concerned about foot problems, as on a self-feed silage system, the cows' feet would be almost continuously exposed to wet slurry, while they were feeding, in the collecting yards or in the passageways in the kennel shed. This seemed bound to soften the sole of the foot and make it vulnerable to infection.

It was awfully difficult to get accurate advice in those first few years, as everyone seemed to have different theories, and it reminded me all over again of the early days with milking parlours. I suppose though that this is inevitable with the introduction of a completely new technique, when there are bound to be a lot of bugs to get out of the system.

Another problem was what to use as bedding, since again, there seemed to be endless theories about the best materials, both for the cow's comfort and also from the aspect of mastitis control. I decided that we would chop straw with an old forage harvester that we had available, as I was already planning a new slurry disposal system which would not be able to cope with long straw. I hoped that the cows would be more comfortable on straw than on some of the other materials that were recommended.

So we took the plunge, opting for kennels for 100, and employing a builder who had already put up a considerable number for other farmers, and taking his advice on the dimensions. With our policy of calving heifers twice a year, we always had quite a few dry cows and in-calvers about, who could be housed in the old yards in the winter. This meant we never had more than 100 cows actually in milk during the winter, and could safely expand to a total of nearly 120 without difficulty. For our particular farm, I was convinced that a twice a year calving policy paid the best dividends, as it created far less pressure both on the herdsman and the youngstock rearer, who had neither a mass of fresh calvers to deal with, nor a lot of new calves to rear. It also meant that we could get by with a very limited range of calf pens, and save on building costs. With a moderate input policy using home-grown concentrate and grain balancer in winter, and grass in summer, we kept up a steady level of production throughout the year with no stress points anywhere in the system. I can perfectly well see the advantage of a tight calving pattern under many conditions, but with our land and facilities, an extended pattern suited us very well, and kept our costs to a minimum. Both on an official recording scheme, and on our management exercises with the students, we came out with margins in the top financial bracket.

As I had feared, the change from yards to kennels did bring a number of problems in its wake, particularly refusal and an increase in lameness. About 5 per cent of the herd initially refused to lie in the cubicles, in spite of all efforts to make them do so. Even if coaxed in, they wouldn't lie in them, but still preferred the wet cold concrete of the silo barn, or even the passageways. Consequently they became increasingly dirty, and obviously rheumatic. Lameness often followed and it was pathetic to see them walking like old people waiting for hip replacements. There was also more lameness both from foot lesions or from apparent slipping on wet surfaces, with consequent pelvic damage, which seemed to throw one of the hind legs out of line. As a fellow sufferer, I thought that I knew exactly how these cows felt, but unlike me they could not ring up their osteopath and get it put right in a couple of visits, or take a walking stick to relieve their pain. On one or two occasions, I did try to apply some heavy massage of the pelvic region, but cows are awfully big animals on which to attempt any manipulation. I seem to remember reading somewhere recently that someone had set up as a cow osteopath and I do hope that I am right in this. I am quite certain that the working life of many valuable cows could be prolonged by some judicious pelvic massage or manipulation – and a lot of pain alleviated, too. I think that there is no more distressing sight than that of watching a herd of cows being turned out to grazing after milking, and seeing about half a dozen poor old ladies – and often not so old either – hobbling along at the back with every step they take obviously hurting them.

It is particularly frustrating to know that in some cases despite all the vet can do the long-term prognosis cannot be good, and that the road will sooner or later lead to the knackers yard, or the cull cow market. I am tempted to wonder sometimes whether the introduction of cubicles was quite the godsend that it appeared to be at the time. It is true that things are better today than they were then, with rubber mattresses and similar comforts being provided, and better design reducing the risk of accidents. But I still read that lameness is one of the biggest problems in dairy farming today, and that was not the case in the days before cubicles and kennels appeared on the scene. It might not be fair to lay the blame entirely on cubicles, as it may also well be a result of keeping far larger numbers of cows together in close proximity on slurry-covered concrete surfaces or, as I suspect in some cases, of very high levels of concentrate feeding. I also read that a number of farmers are now reverting back to straw yards again, even with big herds, and doing it economically through the use of modern equipment and large straw bale spreaders. Perhaps this could be the start of another backward swing of that farming pendulum. I certainly hope so.

Slurry Problems

The other pressing problem by the early '70s was slurry handling, which became acute as the herd increased, especially with 500—600 pigs about the place as well. The slurry from the pigs had to be got out every ten days or so, and during the winter months this caused rutting either on the arable land or the wet grassland. The only available area for spreading in the summer was the dairy paddocks. This was all right as long as the wind was blowing *from* the Oxford direction, for the only people who might suffer then were car drivers on the A40. They were generally travelling at about 70 mph so an occasional whiff of pig slurry was just a short but timely reminder that they were driving through the countryside, and could be no great hardship. But the prevailing wind was from the West, and sometimes it would change to that direction during the day and carry the smell straight into North Oxford, or to the small village that lay between us and the city. For a year or two, there had been references in the local paper about strange smells in North Oxford, but fortunately no one had yet traced them to us! Blame was attached to the local paper factory in the village, which had some kind of waste lagoon behind it. The matter came to a head one hot July afternoon when the wind suddenly got up while the pig herdsman was spreading. This time it took the smell at its most pungent straight over to the paper factory. There it was apparently sucked into the ventilation system, where it was efficiently circulated round the factory and concentrated on the way. I was in the farm office when a furious factory manager, whom I fortunately knew quite well, as he had a holiday house in Cornwall and was a member of my golf club, rang up to say that he had a strike on his hands as his staff were refusing to work until something was done about the smell. He had been in touch with the City Council, the Sanitary Inspector (not yet called an Environmental Health Officer) and several other people besides, and we must stop spreading immediately and never let it happen again, and so on. I pacified him as well as I could, and he quietened down a bit, but it was not long before the Inspector arrived. Fortunately he took a lenient view, as he realised that we did try to ensure that we only spread when the wind was in the right direction and that we could not be responsible if it decided to change direction suddenly. But he did say that we must try to find a long-term solution, as things could not go on as they were, which I had already appreciated.

Naturally there was nothing we could do immediately except to be more careful than usual in picking the right day to spread, but I had been thinking for some time that the obvious way to deal with slurry was to separate it out right from the start into its solid and liquid components, and store them independently. I had read that there were now some separators on the market, so it seemed as if it might be a feasible solution. A comprehensive scheme for the farm would also get over another problem which was beginning to worry me, which was contamination of a Thames backwater some 200 yards away, fed by a ditch which went past the buildings. I was horrified one day to see how much dirty water from the concrete around the farm seemed to be getting through, and I knew that sooner or later the River Authority would have some rude words to say about it.

Though a collection system for all the livestock units would obviously involve a lot of capital expenditure in the first place, once it was in place, it could save a tremendous lot of work throughout the year. It would also prevent damage to soil texture during the winter months, provided, of course, that one could construct a lagoon large enough to take all the liquid during the winter. Here we had a stroke of luck, for the area where it needed to go turned out to have a solid clay base, beneath some eight feet of overlying gravel. We consulted Thames Water and they said that they would have no objection to a lagoon at that site on a clay base, with the sides covered with puddled clay. We could afford to sacrifice quite a large area as it was just a passage field for moving stock down to the flood plain for grazing, and was poorish pasture anyhow. We found that we could get a contractor to dig out a one million gallon lake and puddle the bottom and sides for £5,000. This was a lot better than having to have a smaller lagoon lined with plastic, which might have been prohibitively expensive, and too small a capacity for the full winter storage.

Collecting all the slurry together from the buildings was not going to be so easy, as it involved the dairy, parlour, collecting and dispersal yards, silage area, and kennels, and, in addition, the pig yards, farrowing house, finishing house and all the concrete surfaces around these buildings. The whole area was rather flat, so the channels would not have much fall, but here again we were in luck, because the two buildings which produced the most liquid were at the head of each of the two main channels: the dairy in one case, with all its daily washings, and the pig finishing house at the other. This had a big storage tank built under its loading platform where we could store the very liquid slurry, and then let it out to flush out the main channel if it got filled up at any point with more solid material. It all worked in together very well on the whole, the main channels feeding into a big collection pit situated alongside the separator.

By good fortune, Farrows of Spalding, well known in the irrigation world, had just produced their first separator, and were keen to sell us one, particularly as we would be very happy for them to use us as a guinea pig. Ours was only the third separator the firm had installed, and it was anxious to get as much information as possible on the problems encountered in the field. The separator had rotating arms which squeezed the slurry through a semi-circular screen. The liquid fell through the holes in the screen, and the solids were pushed off by the arms, falling to one side of the separator. But the slurry had to be pumped up to a height of some ten feet with the separator raised on block walls, so as to allow the solids to accumulate beneath it. That was where the trouble started, since our slurry was by no means uniform. It contained chopped straw from the kennels, wads of grass silage from the self-feed area, some straw from the farrowing house, liquids, and washing water. It was asking a lot of a pump to deal with such a combination without getting jammed, particularly at the intake end. We put paddles in the pit to keep the contents as well stirred as possible, but in the early days we did have trouble finding a pump that would do the job reliably. For the first few weeks Farrows' technicians almost lived on the farm, but eventually they solved the problems and the whole system began to work well.

But I *was* concerned at the amount of suspended sediment in the separated liquid, and felt that this should be filtered out or our beautiful lagoon would very quickly get silted up. So I built a 15 foot square chamber under the liquid outflow with three sides composed of old railway sleepers laid on edge and held in place by vertical steel joists. The fourth side was a block wall against the separator. The sleepers could be lifted out with a tractor foreloader. The centre was filled with straw bales, and I hoped that the slurry would percolate through them, leaving the sediment behind. But it just didn't work, as the stuff came out of the separator too fast, never got into the bales and piled up till it overflowed at the top, coming out virtually unchanged. So out came the bales, and in went loose straw in its place, but even that filled up too quickly and overflowed at the top. So there was nothing for it but to abandon the straw, and simply use the chamber as a settling tank. It overflowed at the top when it got too full, and we hoped that sediment would accumulate at the bottom, which could be got out with a tractor scoop.

But very soon came near disaster. Alan Willows, the pig herdsman, had just started the separator and was standing some 12 feet away, when there was a loud explosion. The block wall behind the separator disintegrated, throwing chunks of concrete all over the place. Fortunately the super-structure supporting the separator was between him and the wall, and this took the force of the blast, so he was not hit. But it could have been a very nasty accident. I was to blame for not having the wall properly reinforced but I had not appreciated the very considerable pressure exerted by such a volume of slurry. I was thankful, too, that it had not hap-pened when I was on top of the chamber or I would have been in a bigger mess than I had been when I slipped in the slurry at Brooksby 20 years previously! After that, I gave in, and we discharged the effluent, sediment and all, into the lagoon, and I hear that it has only recently, some 20 years later, had to be dug out, so perhaps I need not have worried about it.

In order to utilise the contents of the lagoon, which was just big enough to hold a winter's supply, we pulled in plastic piping for a distance of some 300 yards down to a series of hydrants strategically placed at the end of each cow graz-ing paddock. An old tractor sited at the reservoir pumped the slurry down to a reel irrigator, so that the whole job was automated, the only work needed being to realign the irrigator when it reached the end of each block. The critical moment arrived when we switched on for the first time, and waited with baited breath, or, perhaps one should say, dilated nostrils, to see whether it stank or not! The reservoir itself had quite quickly acquired a rather pungent odour all of its own, unlike that of conventional slurry. However, it was not too obtrusive about the farm, except in very hot weather. I was reasonably optimistic that everything might be 'alright on the night', so to speak. Thank goodness, it was, as the irrigated liquid only had quite a faint and reasonably innocuous smell, and certainly not one that was likely to cause a rebellion in North Oxford, or even the paper mill. The grass after irrigation certainly had quite a smell to it the following day, but the cows were not put off at all, if we did have to irrigate a bit close to grazing. This was rare, as it was generally applied after first cut silage, to boost the second cut.

The whole works, including the irrigator, cost in the

region of £25,000, which was a lot of money in the early '70s, but it solved at one stroke the whole of our slurry and pollution problems. By the end of the first year of full working, I calculated that we had saved about £800 worth of fertiliser on the grazing and silage block. Taking into account the huge saving in costs that we achieved by not having to spread slurry weekly over the winter, and the mess that it entailed, I was very satisfied that we had made a good investment. This far outweighed occasional problems with the collection system, pumps and separator over the years.

The system is still working now, over 20 years later, so it must have paid some good dividends in the intervening years, and I never regretted the decision to install it. I remain convinced that separation is still basically the best way of dealing with slurry, especially if you are able to flog off the solid material as organic fertiliser to garden centres, as I believe some people have done. I don't quite know why it has not been more widely adopted by dairy farmers.

I have spent sometime on the expansion of our herd and the problems that it brought on our relatively small farm, because they were much the same as those encountered by very many farmers at the time. I have said 'relatively small' because that was what we were for the early '70s, which saw the emergence of some very large herds of 300 cows or more, either in single or multiple units. This created logistical problems in regard to grazing, bulk feeding and slurry disposal. It involved, too, a great increase in parlour size and automation, and the introduction of new concepts of feeding, such as complete diets, and flat rate feeds, which were in complete contrast to the established practice of the individual concise rationing of each cow. It was a very far cry from what I was taught by H. E. Woodman, that rations needed to be adjusted weekly according to previous yield. 'Starch equivalent' had disappeared, and 'metabolisable energy' became the correct term.

Most of these innovations were devoted to the objective of maximising the amount of milk produced per man employed, which meant not only expansion in the number of cows, but also capital investment in machinery and buildings to enable one herdsman to cope without taking on any extra labour. In teaching, one seemed to be forever increasing the number of cows per man that farmers claimed to be employing. In the early '60s this seemed to stick at about 70, when relief labour was included, and additional labour used perhaps for food preparation. But during the '70s, this figure began to increase with more efficient mechanisation. Rapidly rising costs, however, made it very difficult for the smaller farmer, who had not modernised his system earlier, to keep pace. So the number of milk producing farms continued to decline. The EEC Mansholt Plan designed to reduce the number of dairy farmers also accelerated that decline.

This was the era both in dairy and arable farming when the specialist consultant began to come into his own. The MMB had, of course, had its own consultants in the field for 20 years using its Low Cost Recording Scheme. Some of the major companies also ran similar schemes for their farmer clients. This was followed by the introduction of the Daisy Scheme by Reading University, aimed initially at improving herd fertility. The services provided became ever more sophisticated with the introduction of computers. This boosted the consultancy business, as most farmers had neither the time nor the inclination to exploit the advantages that computerised data could provide for herd or whole farm management. In addition, a consultant supplied details of the latest technical advances with which many farmers could not keep abreast. This was especially the case in the arable world where new agrochemicals were continually altering the shape of the playing field. Many large farmers did buy their own computers to do much of this work themselves, but for the medium sized men, it was probably cheaper and more efficient to employ professional services, as we did ourselves.

CHANGES IN ARABLE FARMING

On the arable side of our own farm, we had the extra acreage to deal with, which was devoted entirely to cereals. Oilseed rape had not yet become an attractive crop to grow, though the national acreage was just beginning to creep upwards, especially on the lighter soils where farmers were looking for suitable break crops after the cereal explosion of the '60s. The UK rape acreage was then about 100,000 and it stayed around that figure for a few years, until it took off with the development of suitable market outlets.

We steadily increased our wheat acreage as we got the new land drained and into a better state of fertility, and this increase was mirrored in the national figures. It was caused largely by the introduction of the short strawed varieties, and the arrival of more efficient fungicides and grass herbicides, all of which made autumn sowing less risky, and yields much higher. Plant growth regulators also came onto the market, which reduced the risk of lodging on land of high fertility, and where high input systems were employed. Wheat suddenly began to outyield winter barley by a very considerable margin. Though the variable costs were obviously higher, the gross margins came out strongly in favour of wheat, even on soils previously considered not very suitable for the crop. The turn round was quite dramatic. In 1970, the barley acreage stood at around 5.5 million acres and wheat at about 2.5 million, but by 1980, barley had lost half a million acres, and wheat had gone up to nearly 4.0 million. (Oats had also lost about 400,000 acres as well.) By the middle of the 1980s, the wheat and barley acreages met each other at about 4.75 million acres each.

It was not only the acreage of wheat which increased so strongly, it was also the yields per acre which shot up as well. In the Eastern counties, those damned Jones, with whom we are all supposed to struggle to keep up, could be heard boasting of yields of 4 tons per acre (10 tonnes/ha) which would have been unheard of only ten years before. Personally I made not the slightest attempt to keep up with them, as I knew quite well that our land was not capable of that sort of a yield, and that the sensible thing to do was to try to taper our inputs to the sort of economic yield that we might hope to get in a good year, which was nearer to 3–3¼ tons per acre. We still grew a big acreage of barley as we needed a good tonnage of moist grain for the cattle, and dry grain for the pigs. Though the gross margins for the barley never came out anywhere near those of the wheat, I am quite convinced that we got the difference back in the low feed cost of the animal units, which is where quite a lot of our profits came from. I admit that I did not help our barley costings by sticking to spring sowing. This was for the simple reason that winter sown barley came to harvest in the middle of the only period in the summer when we tried to fit in as many staff holidays as possible. On a mixed farm geared primarily to livestock, the early part of the summer is taken up with two cuts of silage and hay, which usually leaves a short gap for staff to get away in mid July. I suppose that tough 20th century managers would say that considerations like that should not influence the possibility of making larger profits, but I am quite unrepentant. I feel that farm employees are just as entitled to have part of their holidays in the summer months as the rest of us, even if it might mean slightly lower profits.

Towards the end of the decade much less was heard about continuous cereal growing because of the increasing cost of keeping land clean and crops healthy. Profits in general were starting to decline, largely because of the huge increase in costs, which had started with the oil crisis, and then been accentuated by entry to the EEC. Cereal prices just did not keep up with the additional costs. At the same time, the balance had swung back rather more towards livestock. Sheep had become more profitable than for many years previously and mixed farming began to look more attractive. Once again, people began to talk about rotations, with alternative crops and leys being used to help to clean the land and maintain fertility. Oilseed rape had at last begun to take off as a suitable combinable break crop and field pea and bean prices had been helped by joining the Community, and in the former case by the introduction of new varieties. The pendulum definitely began to swing back towards more balanced farming systems, even though the wheat acreage continued to expand, mostly at the expense of barley. The increasing popularity of wheat was due not only to higher yielding varieties, but also to the re-introduction of break crops, which allowed it to be incorporated more freely into the rotation.

In the livestock sector sheep really began to stage their comeback. This was helped by better lamb prices and some restriction on imports from New Zealand, resulting from access to the EEC. By now, too, genetic improvements were percolating down through the industry, and carcase quality was improving, making British lamb more acceptable to markets on the Continent. Higher levels of prolificacy, newly imported breeds, and better husbandry all helped to improve profitability, and to provide an acceptable alternative to beef cattle when utilising leys on arable land.

At the same time, on arable farms in the South, and in a few other areas, the Roadnight system of outdoor pig keeping had begun to spread quite rapidly. This, too, provided farmers with another profitable use for leys on the lighter land farms. The success of the system was due to its low cost in equipment compared to intensive housing and to lower costs in disease control – a factor which was worrying intensive producers. In addition, welfare issues were becoming of concern to the general public as a result of publicity campaigns against the use of sow stalls, farrowing crates, and overcrowded finishing houses. I had myself considered putting in a sow stall house in the early '70s, but quite quickly rejected the idea after seeing a number of examples which filled me with some revulsion. Admittedly, the alternative of housing sows in straw yards had a number of disadvantages, especially that of bullying by aggressive and greedy sows. However, one only has to see the pleasure that a sow gets from partially burying herself in a straw bed and rootling round in it, to appreciate that it must be a much more enjoyable existence than being tied by the neck and having to lie on a bed of insulated concrete with only the next meal to think about. I do accept that one must not assume that animals feel the same way as we do about these things, but pigs do have much in common with humans. This is perfectly clear from the tricks that they like to play on you if you let them, particularly when they are young and you are trying to make them do something that they don't want to. I recall being reported in the press once as saying that pigs have a sense of humour, and I would still stick by that. It can be seen in the eye of any young pig that has got out of its pen, and is challenging you to a game of hide and seek when you try to get it back again.

On the subject of pig comfort, I think that farrowing crates are quite a different proposition to sow stalls, especially as sows get bigger with age. They are generally only confined in them for three weeks or less, and for the first few days are still getting over their farrowing and would not be moving about much wherever they were. There is no doubt whatsoever that a good farrowing crate can save the lives of a proportion of young piglets, or at least protect them from being seriously maimed, quite apart from facilitating fostering and providing a properly controlled environment for the litter. That surely must counterbalance any temporary inconvenience to the sow.

Of course, the outdoor system also has its disadvantages. It is not a pleasant sight seeing a sow forward in pig, sinking up to her belly in mud on a wet, muddy hillside, exposed to a biting East wind. But at least she has a nice warm straw bed in a cosy hut to go back to, and provided that the sows are of the right breeding, they seem to be remarkably resilient and sometimes actually to enjoy the conditions. Their piglets certainly come out of it in an extremely strong and healthy condition. Living as I do in an area containing hundreds of outdoor units, I often feel that it is probably the herdsman who has more to put up with in the middle of a cold, wet winter than his sows!

If there was any one factor more than another which facilitated the revival of rotational farming on the larger arable and mixed farms, it was the electric fence and its various ramifications. Without it, it would not have been possible economically to apply constraints to stock on large hedgeless farms, and this especially applied to sheep. I remember so well travelling back to Oxford one evening with the late Dick Wellesley, that enterprising, lively, pioneering spirit who farmed at Buckland, near Faringdon. He always had something new and exciting up his sleeve which was going to revolutionise farming. He had just got involved in some way with electrified sheep netting, and in his usual enthusiastic way could talk of nothing else. He was quite convinced, and, of course, he was not far wrong, that this was going to make a tremendous impact on the keeping of sheep on arable farms, in the same way that hurdles had done 150 years or more before. Dick's relatively early death was a tragedy that deprived the industry of one of its liveliest thinkers and stimulating personalities.

POLITICAL UPHEAVALS

Farming in the '70s came to be completely dominated by two events. These were destined to change irrevocably the climate in which farmers were to operate in the years ahead, and to impose increasingly restrictive constraints on their ability to farm as they wished. The first was obviously accession to what was then the EEC in 1973, and the second, which many people seem rather to have forgotten about now, was the oil crisis which hit the world at about the same time. In this, Arab oil interests virtually held the world to ransom by forcing up the price of oil to unheard of levels. It was a salutary illustration of how the modern world has become almost completely dependent on the supply of a very few vital commodities for the effective functioning of its commercial activities and its complex infrastructure. Oil is clearly the most essential commodity today, and, on a more local basis, it is electricity. In the past, I suppose that coal was almost as important as oil is today. Certainly I have pleasant memories of the coal and general strike of 1926 which caused tremendous disruption, and gave me an extra 10 days Easter holiday, as I could not travel back to boarding school!

In the early '70s, there was not the range of supplies of oil that there are today and the OPEC countries were able, by cutting production, to apply the screws needed to force up the price. This had an almost immediate effect on the price of virtually every commodity needed by farmers – fertilisers, feeding stuffs, machinery – the lot, in fact. Quite suddenly costs of production began to escalate, but because of the long-term nature of agricultural processes, there was always a considerable time lag before any compensating price rise for the finished product could be obtained. Even when the rise came it was not enough to make up for the increased costs. The result was a severe cost/price squeeze on farmers which continued inexorably throughout the '70s as inflation rose to unprecendented heights.

All this began to happen at much the same time as Britain entered the EEC, when farming had to adjust to completely different systems of support. The old system had been designed to keep food costs down by paying relatively low prices to the farmer. These were supplemented by deficiency payments, or by direct grants and subsidies. This had to be abandoned in favour of the Community system by which farmers got their return from the market. Support for the main products was provided by Intervention Boards which took surpluses off the market when prices dropped below a prescribed level. The cost of doing this was financed out of levies imposed on food imports from countries outside the Community. At the same time a complicated system of monetary compensatory allowances was devised, in an attempt to get over the problem of different currency rates between Community countries. Additional support was provided by grants for special cases, the best-known being that for the less favoured areas (LFAs). Under this heading Britain had a special claim for the hill areas of the West and North.

The immediate effect of entry was to raise the price of cereals, particularly, with modest price increases for other commodities, and it was not long before overproduction began to fill the Intervention stores with grain, and milk products. The effects of entering the Community and pushing up the price of food were to some extent masked by the inflation caused by the oil crisis, and the general public probably never appreciated the actual impacts of entry to the Community.

By the end of the decade, Intervention stores were full, and perishable foods, such as butter, were having to be sold off at very low prices on the world market. The difference in price had to be made up out of Common Agricultural Policy funds, and eventually from direct Community grants to keep the CAP in business.

Accession to the EEC meant essentially two things to British agriculture. Firstly, control of prices and policy shifted from Whitehall to Brussels, and the annual bargaining between the Government and farming interests at the February price review disappeared. In effect, it was now the Government that was left to fight the British corner in Brussels, and farmers could only try to influence things through the European farmers organisations, which were certainly not so effective as the NFU's had been under the old system. Secondly, British farming had now become only quite a small part of the total Community farm, and instead of being in a deficiency production situation in many commodities, it was moving rapidly into an overall surplus regime. This was caused by overproduction in some countries like France, where the strong farming lobby had ensured prices high enough to keep inefficient farmers in business. The dairy sector was a typical example. Britain on entry was only some 80 per cent self-sufficient in milk and dairy products, but on becoming part of the whole unit, which included Holland and Denmark, we quite quickly reached the point of massive overproduction. This resulted eventually, in the early '80s, in the imposition of milk quotas, although in Britain we were still not self-sufficient. The general result was that by the end of the '70s, the previous rate of expansion had slowed right down, and even ceased in some sectors, and farmers incomes in real terms were considerably lower than they had been at the beginning of the decade. Investment had also slowed down markedly, and bank borrowings increased significantly, while a lack of confidence and a feeling of uncertainty had begun to pervade the industry.

It is, of course, quite impossible to say what would have happened if Britain had not joined the Community at that time. There would undoubtedly have been a serious pressure on prices, and oil inflation would have accentuated the cost/price squeeze, as the Government tried to reduce the subsidy bill and force the industry into higher levels of efficiency. Most people seem to think that it would have been worse if we had not gone in than it subsequently became under the Community regime. Perhaps it might have been for a year or two, when the higher prices from the EEC cushioned farmers, but in the long term I am not at all sure that Britain really benefited from the change, since it lost any chances of exerting pressure directly on policy makers (hence the frustration that we feel today). I was always extremely critical of the whole European Community concept, and especially so with regard to agriculture. I could never believe that it would be possible to formulate a fair and workable policy which would encompass, on the one hand, a very efficient 300 cow dairy farmer in Cheshire and the small Italian peasant farmer that I used to watch cutting grass by hand for his six cows, when holidaying on Lake Como. The same could be said of a 1,000 acre Lincolnshire farmer in relation to a small cereal grower in central France. I did not trust the euphoric statements made by the politicans, that we had a market of 150 million people waiting to buy our goods, when it was perfectly obvious that the French and the Germans and the Dutch couldn't wait to exploit our market of 57 million. I was never one to expect either that the French or Germans would play the game by the Queensberry rules. The following letter which appeared in *The Times* on 10 June 1971 should exonerate me of the charge of being wise after the event.

Farming under the EEC
From Mr M. H. R. Soper

Sir,

Professor Denman's article (June 3rd) is indeed timely in drawing attention to the effect that joining the EEC could have on the price of land, and the future prospects for those who wish to enter farming. It is, in fact, astonishing that the agricultural industry in general has shown so little awareness of the dangers inherent in joining the Community. It has, it would seem, been lured by the twin siren songs of the politicians to the effect that cereal prices will rise dramatically on entry, and that Europe is prepared to pay high prices for beef. It is probably true that for a few short years British cereal growers will be batting on a good wicket, but it should be remembered that 68% of the gross sales off British farms are of animal products, that most of these rely heavily on grain as a feedingstuff, and that the Community is already virtually self-sufficient in several of them, or would be so if Denmark and Eire were to enter with us. Because of this, a considerable proportion of our farmers will face greatly increased costs through a rise in the grain price of some 50%, with at the same time, little prospect of a commensurate price increase for their products. Their support system will have been abandoned for an untried system which can apparently allow the Government to fix a bottom to the market as low as it pleases, and the large sum of money collected in import levies, instead of being devoted to the problem areas of British agriculture such as hill or marginal land will have to be paid into the Common Fund, where it will presumably be used to pay for the Mansholt Plan, which will make Continental farmers more competitive with our own.

But this is not all, for the sharp rise in food costs (and whatever the politicans and theoretical economists may say, there are *bound* to be very large rises) will have a snowball effect on the cost of the farmer's raw materials. Every person handling fuel, fertilisers, machinery and feedingstuffs will demand higher wages because of the rise in his own food and living costs, and these wage increases will inevitably be added to the cost of such raw materials. This compound increase in costs will bear extremely heavily on the farmer, for he is engaged in a long-term production process in which his costs increase many months before his returns can increase – if indeed they ever do to the same extent.

If Britain does enter the EEC the farmer is going to be caught up in an inflationary squeeze of very serious proportions, and the agricultural industry of this country, instead of being one of the best equipped and most progressive in the world could well be reduced in the course of a few years to the poor relation conditions of the 1930s. This will, of course, for other reasons apply even more forcibly to the horticultural industry. It is sincerely to be hoped that farmers in general will not be lured into exchanging their birthright for a mess of pottage, or should one say pôtage, by the promises of a rise in grain prices, which in real terms can only be of a temporary nature, and for a somewhat dubious future market for their beef.

Yours faithfully
M. H. R. Soper
Beech Cottage
Marcham, Abingdon.

Reading that again after some 25 years, I don't think that there is much that I would wish to retract. When it came to voting time, I drove around with a 'Keep Britain out' sticker on the car, but it didn't affect the issue, of course, and we were all led like lambs to the Brussels slaughterhouse. The forecast in my letter about increases in costs came about soon enough, even though they were also influenced by the oil crisis.

STRUCTURAL CHANGES IN THE INDUSTRY

Brief mention has already been made about changes in other aspects of the industry over the years, and one of these is perhaps best discussed in the context of the '70s, though it had been influencing farming for a long time, and is still doing so today – land ownership. Since the end of the war, a continual and almost insidious change had been taking place in the traditional landlord and tenant system. This had persisted since the enclosures, and probably reached its zenith in the prosperous farming times of the mid-19th century. Its great advantage was that it shared out the capital requirements for farming between the two partners, who both had quite clearly defined responsibilities. When farming was prosperous it worked well, with landlords willing to invest capital in improvements on their farms for the benefit of the tenants, who were themselves enabled to farm more profitably and stand higher rental charges. But when farming became less profitable, or external factors such as high taxation or difficult economic conditions curtailed the ability of landlords to maintain investment in their estates, the system did not work so well. Neither partner was then able to play their full part. In the period between the wars, both external factors were operating and there was a general decline in the standards of management, especially where small estates and poorer quality farms were involved. The sort of conditions that we encountered on taking over our farm at Oxford were by no means atypical, for I had met with a number of farms and small estates in similar circumstances during the War Years in Surrey.

Since the Second World War, estate duties, taxation, periodic business recessions and so on have all conspired to accelerate the demise of the system, so that in the 30 years after 1946, the proportion of rented farms fell from approximately 75% to some 45%, and the figure today stands at about 35%. Direct owner occupation increased commensurately, though in some areas, such as the Eastern counties, the land has moved into company rather than individual ownership, so that the actual farming is done by managers rather than by sitting owners. In the smaller farming areas direct owner occupation is still the norm. In effect, of course, the banks and other financial institutions have taken the place of the landlords through providing the money for purchase, though without any of the landlord's responsibilities. This is one of the principal reasons for the huge increase in the indebtedness of the agricultural industry to financial institutions that has been such a feature over the past 20 years.

As land values escalated rapidly in the '70s, and some City institutions moved into farming directly, the competition for land became intense, reaching the point where it became almost impossible for young people to acquire land to start farming. Sadly, that situation still holds today, even though some of the City money, which is notoriously fickle, has by now moved out again.

Another factor which was highly significant in its effect on farming was the passing of legislation by the socialist Government in 1977, which gave security of tenure to certain members of a tenant's family on his death. I well remember the furore that this created at the Oxford Farming Conference that year, since it was obvious that it would impose impossible restrictions on landlords. It did have a disastrous effect on the number of farms that became available for letting, since landlords were clearly unwilling to create new tenancies with relatively young people, as it might mean tying up their farms for up to 100 years by the time that the new tenant's son, who would have inherited it, eventually died. Landlords therefore took the land in hand instead of letting it, and farmed it direct or through arrangements with local farmers or contractors. Either way, it meant fewer farms available for young people to get into the industry on a tenancy basis. The new legislation liberalising the tenancy laws, coming into effect in 1995, must be beneficial but whether it will provide more farms for young entrants is doubtful.

The main fall-out in land ownership in the '70s was in the small to medium estate sector, so that today a high proportion of the tenanted farms still available is owned either by very large historic estates or by institutions of various kinds. It seems most unlikely that the clock will be put back, especially as many county councils are now selling their smallholdings, which were one of the traditional routes for young entrants. The number of farms to let has been greatly reduced, too, by the consolidation of larger estates. Our own Wytham Estate is a typical example of this, whereby the original seven farms in 1946 are now down to two through amalgamation. When farms do come on the market today, they are often bought by neighbouring farmers, which reduces their numbers still further. Ultimately, it could evolve to a situation where perhaps some 20 per cent of the land would be in the tenanted sector, as I would not anticipate that either institutional ownership or very large estates would disappear altogether.

ENVIRONMENTAL ISSUES

Another factor which began to assume considerable significance in farming circles during the '70s was the growth of the 'green' movement in its different forms. It had, I suppose, really started much earlier in a minor form, even before the Second World War, with Sir Albert Howard, and Lady Eve Balfour leading a crusade on behalf of organic farming. Many of the myths about the sanctity of so-called 'natural' methods of production, and the perceived iniquity of using chemicals of any kind when growing crops or feeding animals date from that time. But in the need for all out food production during the war, and into the early years of peace, such theories did not cut much ice. The general public certainly didn't worry at all where its food came from, as long as there was enough of it at a reasonable price, which in spite of all the current propaganda, I believe still to be the case today.

But with the expansion of farming in the late '50s and '60s, with its removal of hedges and felling of trees, the ploughing up of old pastures, the loss of wildlife habitats, and the use of increasingly large amounts of chemicals, many people suddenly became alarmed at the destruction of the natural environment that was taking place. At much the same time, too, animal welfare issues suddenly began to be discussed in the media, starting with concern at battery cages and broiler houses, and moving on from there to pig buildings and veal units. This led to the setting up of the Brambell Committee, which finally reported in 1965. The resultant codes of practice for animal welfare were accepted by the industry with some relief, but have never satisfied the more vociferous critics. So a number of quite different environmental issues all came together at about the same time, most of which were directly related to current farming practices, though pollution by other sectors of industry quite rightly began to attract much greater attention as well.

As far as farming was concerned, none of this anxiety was wrong in principle, for in the rush of farm expansion, there were undoubtedly abuses and excesses in some areas of production. These caused damage both to animals, and to the environment in general. But what *was* wrong was the implications put out by propagandists that *all* farmers were involved in these practices, and that the whole of the natural environment was being put at risk. They gave the impression that Britain was being turned into some kind of Middle West prairie, where wildlife was obliterated, and all farm animals were kept in appalling conditions and subjected to cruel abuses. Unfortunately, the media, with its insatiable appetite for sensational stories, and reluctance to transmit good news, very often latched on to the bad cases, but failed to present the other side of the coin.

By the '70s, when many of the more thoughtful farmers had begun to get very concerned about the image of farming, and Farming and Wildlife Advisory Groups (FWAG) were born, it was almost too late to do much about correcting the impression that had been implanted in the mind of the public. The important thing was to hold the fort, to encourage farmers to abandon the more questionable practices, and to adopt methods which were environmentally friendly, while counteracting the more extreme accusations.

I have to admit that, until the early '70s, I had not given much thought to actual environmental problems. That was possibly because I had never considered pulling out a hedge (except in some necessary cases during the war), and because I had always, as I have said elsewhere, been a moderate input/moderate output type of farmer, both with livestock and crops. Then I had a visit one day from Jim Hall, a retired Eastern counties farmer, who was campaigning on behalf of environmental protection from the more extreme abuses. He came to speak to our students' club, and pressed me in my capacity of Chairman of the City and Guilds Committee to include environmental issues in our syllabuses, which were undergoing their second revision at the time. I am afraid that we did not accede to that request, since we were already trying to squeeze a quart's worth of information into a pint pot of syllabus. My working party thought, in any case, that this was more a matter for management than for craft students at the first or second stage of a day release course. But I was impressed with his arguments, particularly as I had quite recently got involved with the Association of Agriculture, which operated in a somewhat similar field – relations between agriculture and the general public. The FWAG movement really established itself in the '70s, and by the end of the decade had done much good work in alerting farmers to the dangers of extreme practices. It had started positively to mend some of the fences with many of the countryside organisations – work that has continued up to the present time.

The Association of Agriculture was an educational charity, which had been set up in 1947 in the aftermath of the war by a group of prominent agriculturists and industrialists, with the help of bodies such as the NFU and the CLA. Its purpose was to promote a better understanding of the countryside and farming, and to provide information to the public about the sources of its food, and how it was produced. It was, too, a further attempt to ensure that agriculture should not again become a lame-duck industry, producing only a small proportion of the nation's food. 1947 was a good time to launch such a body since farmers were more popular then than they had ever been (or have been since!). Commercial firms were prepared to put in some funds to start it off, though, sadly, many of them subsequently fell by the wayside, and the initial funding was never enough to build up a proper reserve fund to provide income in the future. It had also got off to rather a bad start with a first director with grandiose ideas, who spent too much right at the start. My old friend, Sandy Hay, had then been brought in to pull it round.

I first became involved in 1968, when Sandy asked me to give a talk about modern farming and technical developments in agriculture at a Teachers' Conference in Bristol. In getting the Association straight Sandy had decided to concentrate the limited resources primarily on work in schools, and in providing general information about farming to the public. I was very impressed with the meeting, and the enthusiasm of the teachers, who seemed very keen to learn more about agriculture. I also admired the extremely efficient organisation of the whole event by Joan Bostock, Sandy

Hay's assistant, so I became a member. In addition to a Council and an Executive Committee, there was an Education Sub-committee, responsible for publications, organising courses, arranging farm visits for schools, and so on. This had been chaired for many years by none other than our old friend Sir James Scott Watson, who had retired some years before, to be succeeded by Bill Garnett of the Norfolk College of Agriculture. But he wanted to resign, so I quite soon found myself persuaded by Joan and Sandy to take it on in his place. This was the third occasion on which I was called on to take on responsibilities previously shouldered by Scott Watson, who I held in the highest regard.

There was a great deal to do, but very little money to do it with, since the Association was entirely dependent on voluntary subscriptions from commercial companies and a few agricultural organisations. Joan succeeded Sandy as General Secretary, and, as is the case with so many charitable bodies, she had to spend almost half her time trying to raise money to keep it going. There was a small membership, consisting mainly of teachers, and a few dedicated farmers. We never seemed to be able to make a breakthrough in the farming sector, except for valued support from the NFU, who referred many of the requests for advice that they received on educational matters to us. Joan was an extremely capable administrator, with a good knowledge of agriculture, and of the education system. She had worked as assistant to John Green at the BBC in the Agricultural Unit throughout the war, and had then been with the British Council in Vienna for some years. She had acquired considerable skills both in the literary field and in public relations, which were quite invaluable in running the Association. She had also established a network of contacts across the country which was of great value in running the school farm visits scheme, through which any school wishing to visit a farm could be put in touch with a suitable one in its locality.

When I took over as Chairman of the Education Committee, I felt that we needed to have a lot more literature available on agricultural topics, which would not be too technical, but would provide the sort of information on farming issues that a teacher might need for a geography or biology syllabus. We had very little literature of our own available, except for a range of farming case studies which Joan had written over the years, some of which were now getting out of date. Publications might also help to bring in a few pennies, and get our name better known. So we started off with a series of what we called information sheets. These varied from 3–8 photocopied sheets, which covered a wide range of topics from the purely technical such as the rotation of crops, to political issues, such as the Common Agricultural Policy of the EEC. I wrote quite a few of these myself, and we commissioned experts to write those that we could not handle ourselves.

One of the objectives in producing these sheets was to try to counteract some of the more outrageous statements made in the literature produced by anti-farming lobbies, which tended to target teachers when trying to put across their views. Animal rights issues were a case in point. We produced 24 of these sheets, which clearly filled a gap, as they sold well. It then became clear that some teachers were looking for rather more detailed material, but still of not too technical a nature. This was rather later in the '70s, as teaching theory began to change, with a swing away from the memorising of facts towards project work and making pupils find out things for themselves. Both in biology and in geography, it was realised that the farm could act as an excellent outdoor laboratory, but many teachers, mostly coming from urban backgrounds, needed sources of quite simple background information in order to be able to make the best use of farm visits with their pupils.

It was quite essential that this information should be as factual and unbiased as possible. I was well placed to assess this need, as I had for some years been the Chairman of the Environmental Studies Committee of the Associated Examining Board, one of the big five GCE examining boards. This committee had grown out of an agricultural science committee, which had outlived its usefulness. We had converted it into an applied biology syllabus committee, which was much more in keeping with modern teaching methods in biology. It relied quite heavily on the use of the countryside and the farm for the teaching of biological principles.

To fill the gap for teachers requiring more information in depth, we produced a short series of booklets covering different farming enterprises, and we were fortunate in obtaining a grant from the Frank Parkinson Trust to enable us to print these and market them ourselves. We launched the series in 1978 with a 48-page booklet, entitled *British Agriculture Today*, which was packed with facts and figures about the industry. It went into three editions over the next eight years, with a 2,000 print run, so it clearly met a need. Most of the copies went to schools and colleges. We followed that one with six others on cereals, pigs, poultry, dairying, soils and grassland over the next few years. Of these, I wrote five myself, commissioning those on pigs and poultry from teachers at agricultural colleges. In any writing that I did for the Association, I would always try to have in mind a teacher, or an interested member of the public who needed straight factual information in a concise form. I tried to put the case of the good responsible farmer, who cared for his land and its future, and attempted to correct, with facts, the misstatements made by factions with axes to grind.

One of the problems in producing a series like this is to keep them up-to-date in the case of such a rapidly changing industry as agriculture. I managed to do this by producing a new photocopied sheet each year of updated tables and graphs. This was slipped into each booklet as it was sold.

This work with the Association of Agriculture continued for some years after I left Oxford, when I was pushed up to Chair the Executive Committee.

FURTHER DEVELOPMENTS IN EDUCATION

Development work on the City and Guilds courses continued unabated throughout the '70s with a complete revision of all syllabuses to give them a more practical slant. Compulsory practical tasks were incorporated, and marks from practical tests contributed to the final assessment. This change was accompanied by an increasing amount of continuous coursework assessment, which was then becoming fashionable in education. The problem with coursework assessment with external examinations is to ensure comparability of standards between different teaching centres, and this did for the first year or so present us with a few problems. Foreseeing this we started off with the marks for this component set at a fairly small proportion (20%) of the final total. But as centres became more experienced, and we were able to bring those which were abnormal into line, we increased this to 30% and then to 50%. This meant rather less reliance on the written theoretical exams, which many good practical students found hard to deal with. That we were able to do this, was a measure of the pace at which local teaching centres had been able to equip themselves, through their education authorities, to deal more effectively with the requirements for practical instruction. Naturally this varied up and down the country, but, on the whole, it was very commendable.

Nothing stands still in education, as in farming, and no sooner had we established what we thought to be a good balance between theory and practice than along came two other developments. The first of these was the report of the Hudson Committee on agricultural education below degree level. It recommended one examining authority for the part-time courses and the one year college courses and that examinations at diploma level should come under the Business and Technician Examinations Council (BTEC). It also suggested that Farm Institutes should become colleges (which some had already).

The recommendations took several years to implement, but no sooner were most of them in place in the '80s than the Thatcher Government came out with its proposals for National Vocational Qualifications (NVQs) for all sectors of industry, under a completely new council. This laid down six levels for practical qualifications, from level 1 at early craft standard, to level 6 at professional management level. This meant a complete reassessment of the role and structure of the City and Guilds and also the regional examining bodies schemes, and their methods of practical assessment. As it turned out, agriculture was about the most forward of all industries at levels 1 and 2, since many of the requirements of the new scheme had already been incorporated in our syllabuses. But I decided that this was the point at which it was time for me to go. I had done some 15 years as chairman of the City and Guilds committee, during which time we had achieved a tremendous amount in bringing part-time education in agriculture into line with that for other industries, and helped to raise the standards of teaching through the medium of our examinations system. It was clearly the time at which someone else should come in and guide the work forward (if it *was* going to be forward, about which I had grave doubts) under the NVQ regime. I did, however, stay on as Chief Assessor of the National Certificate Board, which was then absorbed into the National Examinations Board for Agriculture, Horticulture and the Allied Industries (NEBAHAI) as recommended by the Hudson Committee. I suppose that all this change was, in fact, progress, though I felt that much of it was politically motivated, and smacked of 'change for change's sake'.

Strangely enough in the early '70s I had been involved in the concept of vocational education on behalf of City and Guilds, though it was in rather a different form to that enshrined in the NVQ scheme. Olive Foss at the Institute had heard of a move in certain parts of the United States towards the inclusion of vocational education in the school curriculum for 14–16 year olds. She wanted to know more about it, and, hearing that I was going over to California in 1974, asked me to look into it, and report back to her.

The Institute provided me with some names and addresses in three different areas: one quite close to Los Angeles, one near St Louis Obispo, and one in Sacramento, the state capital, which lies some 100 miles inland from San Francisco. The last of these proved to be the best contact of all, as the Director of Education there referred me to a large school in Modesto, near to the centre of the fruit and raisin growing area, inland south of Sacramento. Quite apart from education, this turned out to be a fascinating place from the horticultural point of view, with huge areas planted with vines, and the grapes laid out to dry in the sun. There were also vast orchards of almond and peach trees – the main home, in fact, of Del Monte fruit and Sunmaid raisins.

The local Director of Education – a real live wire – was completely dedicated to the principle of introducing some craft education into the school system, and I spent the better part of a day with him visiting what was clearly his showpiece of a school – and an impressive showpiece it was, too.

The thinking behind the American development was that large numbers of boys and girls were coming out of school at 16, with no conception of the world of work, or any idea of what they really wanted to do. Consequently industry, and employers in general, were spending huge amounts of time and money in the elementary training of new employees. After a short time these teenagers decided that it was not for them, and drifted off into something else where the same thing happened again. I was given some fairly horrifying statistics, I remember, for the number of job changes that school leavers who were not progressing into higher education made before they settled into a permanent job – if they ever did. And that was another reason for experimenting with the school courses, for the education and social service authorities were becoming very seriously alarmed at the increase in the number of drop-outs. These seemed content to have no jobs at all, but to rely on social security for their maintenance, and many of them ultimately drifted into the criminal or drug culture society. It was hoped that if a proportion of them could be actively motivated before they left school, it might go some way, at least, to ameliorating the problem.

Where it had become fully established the system seemed to be working well, as exemplified by the school in Modesto. But it had meant the investment of quite large amounts of capital in the provision of workshops and similar facilities. This problem was partially met by the involvement of local industrial firms, whereby they provided facilities for training on a day release basis. Other firms who could not provide such facilities were prepared to contribute funds to help the school to do so instead, or to provide equipment, such as computers.

There was a very active management committee for the scheme, on which there sat representatives of local industry and commerce, as well as parents and teachers. There were seven options for boys, which included subjects like automobile engineering, carpentry, electronics, television engineering, agriculture, computers and catering. For girls, there was catering, secretarial work, home economics, computers, and design and dressmaking. At fourteen, pupils were allowed to 'taste' different courses to begin with, before deciding on a definite one, but fifteen- to sixteen-year-olds had to opt for a definite course and stick to it. Each pupil spent one day a week on the vocational course, with instruction being both on theory and practice. On the agricultural course, some time was spent on actual farms, mainly working with livestock. Those on the catering course provided the meals for the school canteen for the day, under supervision, of course, which meant meals for some 1,500 pupils and staff – and an excellent three course lunch we had that day, too. Some of the other units such as automobile engineering and the television section undertook outside repair work under contract, and were thus able to generate some income which was devoted to paying for further equipment.

I was very impressed with this particular set-up, which was well in advance of the others that I had looked at. The involvement of the local community and industry was particularly impressive and this seemed to be a very essential component of any such scheme. The staff claimed that it was also good training for those who were going on to further education and to university, and that the one day out did not seriously impede their academic work. I did not quite see how this particular type of vocational education could be adapted to fit into our school system in Britain. It was also hard to envisage where the money would come from to provide the facilities, in view of what was already spent on Technical Colleges catering for post-school day release and evening classes. One could not see many teachers welcoming a development which took their pupils out of regular school work for one day a week. I could not quite see, either, where City and Guilds might find a niche in any examination scheme, if such courses were to be started.

On the flight home, I compiled a report for Olive Foss, which recommended that there were some important issues here which might well be worth investigating in greater depth. The Institute then sent quite a senior member of its staff for a three months' study tour, and the result of his report was the production of the City and Guilds' Foundation Course syllabuses. These combined both elements of general education, such as numeracy and literacy, with practical facility over a range of technical subjects. They had quite a large uptake from schools whose curriculums contained an element of practical training in some subjects, and from colleges dealing with first year post-school candidates. These pupils often had limited academic qualifications and found the three R's component of the Foundation Course particularly helpful.

Finally, as a last reference to City and Guilds, I have much cause to be grateful to the Institute for giving me the opportunities over the years to gain such an insight into many facets of technical education, over and above my own purely agricultural commitments. I do count it as a great honour to have been elected to the Honorary Fellowship of the Institute – a comparatively rare award for an 'outsider', I understand. This was accorded to me soon after my retirement from the National Certificate in Agriculture Board in the mid '80s.

MORE EXTERNAL ACTIVITIES

On a personal basis, other extremely stimulating commitments came my way during the '70s – first and foremost was an invitation to become a member of the Advisory Committee of what was then the Institute for Research in Animal Diseases (IRAD) at Compton. I had got to know the Institute well over the years, ever since it was set up, in fact, under Dr Gordon soon after the war. I used to take parties of students there at quite frequent intervals. The Director by the '70s was Jack Payne, well known for his work on metabolic profiles in cattle. He had decided that, with a 2,000-odd acre farm, it might be sensible to have a member on his committee involved in practical farming, in addition to the ADAS representative, who at the time was Ted Bullen, the head of the Experimental Husbandry Farms section at the Ministry. All the other members were distinguished scientists or veterinarians, whose presence was obviously essential since so much of the Institute's work was concerned with pure scientific research.

The farm had basically two functions to perform: firstly to supply large numbers of animals for the different research departments, and secondly to provide field facilities for testing out in practice the results emanating from the basic research. The farm, situated high up on the Berkshire Downs, and surrounded mainly by large arable farming units was well placed to produce healthy, disease-free stock. At the time there were three dairy units, a sheep flock, and a pig unit, but many of the experimental animals were housed in the very large compound, containing over a dozen experimental buildings. These were all serviced as far as the cattle were concerned from very large concrete block silos. The silos were filled largely with lucerne and ley grass silage, or with maize silage. These were labour intensive as far as the farm staff were concerned. Though all the animals used in experimental work were costed in at market prices, the farm still did not seem to be paying its way as well as it should have been, in view of its large arable acreage and conditions at the time. The yields were fairly modest by modern standards.

The work of the Advisory Committee was largely concerned with research priorities, progress reviews of the large number of experiments and advising the Director on the expansion or curtailment of current programmes. Towards the end of my time with the Committee we were also involved with the construction of a new, very high security compound for the study of very infectious diseases, where complete containment was essential.

Shortly before I was due to come off the main Committee in 1978, Jack Payne persuaded the Research Council to allow him to set up a small Farm Advisory Group. This was because he was still concerned that the farm did not seem to be contributing financially as much as he thought it should. There was also some talk, I believe, that as research money was getting much tighter, there might be a case for reducing the size of the farm by selling some of it off. This would clearly have been a very short-sighted policy, but to counter it Jack's hand would be much strengthened if the farm could be shown to be paying its way. The members of the group consisted of Ted Bullen, Chris Tozer, the well-known Hampshire farmer, Malcolm Stansfield from Reading University, whose estate at Churn adjoined Compton, and myself. I was asked to chair the group.

It had been set up rather in the nick of time, because retrenchment was already in the wind, and it was not to be long before the Director was instructed to make redundancies. This involved not only scientific staff, but farm staff as well, so some reorganisation was clearly going to be needed. The main problem that we had to face was that most of the buildings had been put up just after the war, and had had practically no money spent on them for 30 years. Many of them were obsolete, having been planned in an era of man power, rather than machine power, and were unsuitable for mechanical handling of materials. The pig unit was very much out of date, and one of the three dairy units was just not worth modernising. Fortunately, some money had been found to expand one of the other dairy units for a new large-scale experiment, so it was possible to close the obsolete one completely and save staff.

There was also at the time a strong demand from the research side for an increased number of young disease-free pigs, so funds could be found for new farrowing accommodation, and a start made on improving the finishing pens. Quite a lot of the arable equipment had to be updated to larger machines. This enabled work to be done more speedily with a lower labour input and better timeliness, and cereal yields soon began to improve.

Once such things had been sorted out, the corner was turned and income began to increase. Greater income enabled more improvements to be carried out, and these in turn increased profitability. As is so often the case, once a beneficial spiral can be created it gathers speed, generating its own momentum and recharging its own batteries. I like to think that our small group provided the initial impetus, and certainly turnover began to increase rapidly. One felt that the farm was now at least justifying its existence as a farm, and not just as a supplier of animals to a research unit. I chaired the group for several years into the '80s, and then unfortunately felt compelled to resign over a completely different matter. This was the closure of the Letcombe Research Station near Wantage, and the threatened imminent closure of the Weed Research Organisation at Begbroke by the Research Council.

Both of these stations had had their origin in our own Department at Oxford. The former had been set up primarily as a radio biological unit under Scott Russell from our staff, and the latter had sprung from Geoffrey Blackman's weed research team, when it outgrew the facilities which could be provided on the Field Station at Wytham. Letcombe, under Scott Russell, had developed into a very good small research station. It did first-class work on soil/plant relationships, nutrient uptake, root development, and the effect of cultivations on soil structure. It had a small very well-integrated staff, and of all the Institutes that we visited with our students, it received the highest marks from them for interest and quality of interpretation. It seemed to me dreadfully short-sighted to close it down and break up the teams just because it was small. I felt much the same about the Weed Research Organisation at Begbroke, which was doing excellent work in its own field, and provided independent assessment of the wide range of herbicides coming onto the market, quite apart from anything else. It was proposed to expand Long Ashton at Bristol into an arable crops research institute to do some of their work. This seemed illogical when all the main arable areas of the country were in the East and the Midlands. Another part of the proposal was to expand Rothamsted, but this was already a large institute which appeared at the time rather to be living on its past historical achievements. I went public with my protests with a letter in *The Times* (Yes, another one, but that is the last resort of the Englishman who wishes to protest against authority!!). I then found myself engaged in a face to face discussion on the BBC *Farming Today* programme with the Earl of Selborne, who was then the Chairman of the Research Council.

In view of that and of the nasty things that I was saying in public about the AFRC's policy, I did not feel that morally I could remain Chairman of one of its committees, however insignificant it might be in the overall context. So I sent in a letter of resignation before the broadcast was recorded. It was probably a good thing as it gave Compton a chance to bring in some new blood, but I was unhappy about it at the time, as it was one of the best and most interesting little committees on which I have ever sat. Of course, the protests which came from a number of different quarters were of no avail, as all the Research Councils were being squeezed very hard by the Treasury, and the blame really lay at the door of the Government. This was only the start, unfortunately, since the next eight years were to see the closure of a number of other important Institutes, including Grassland Research at Hurley and the NIRD at Shinfield, which was particularly sad in view of its distinguished past. Though the debate with John Selborne was a bit acrimonious at times, I am glad to say that the hatchet was buried not long afterwards in connection with an altogether different activity. The amount of money wasted by the Government in setting up expensive research stations and then closing them down again after quite a short time through lack of funding was disgraceful.

It was in 1970 that an event occurred which was to alter the course of my later life to a considerable extent, and is still doing so. Quite out of the blue I got a letter in late 1969 from the Secretary of the RASE. He wrote that three of the Royal Agricultural Societies – the RASE itself, the Royal Highland of Scotland and the Royal Welsh – were promoting an Awards Scheme, to recognise individuals who were making an active contribution to the agricultural industry. They had set up a Council, under the Chairmanship of the late Everard Hosking of the RASE to promote and run it. There were to be two grades to the Award: Fellowship and Associateship. It was intended to elect Associates from younger people in the various branches of agriculture, who were already beginning to make a mark in the profession. If that promise was fulfilled they could be elected some years later to the Fellowship. In order to get the scheme off the ground, the Council had selected some 70 persons from nominations put forward in the three countries, to whom it was proposed to offer a Fellowship by direct election. These would constitute a foundation body with representatives on the Council and they would play a large part in the selection and approval of candidates for Associateship. My name had come out of the hat, so to speak, and I was asked if I would accept a Fellowship.

I thought that it sounded an excellent scheme for raising the status of agriculture in a professional way. The industry had never, till the war, been regarded as a professional undertaking, but it now had many distinguished practitioners, both in the world of practical farming and also in the ancillary sectors, which provided so much support for the farmer in the field. There had never been a body which represented the industry as a whole, in the way that the BMA for example represented the medical profession, or the CBI, the industrialists. Yet the industry by its size and contribution to the wealth of the nation was as large as any other single profession or body of producers. Might this turn out to be a means of promoting a kind of professional body which because of its distinction could represent the views of the industry as a whole? It seemed to me that something like that could develop in due course, if the scheme progressed and expanded, though clearly it was not something for the present.

Anyhow, I was suitably flattered that someone should have thought fit to put my name forward for nomination. We were duly elected and received our Certificates of Fellowship at a special meeting at Stoneleigh, from the Dowager Duchess of Gloucester, who was the President of the RASE that year.

That was fine, but the next few years were to show that the scheme had a fatal flaw in it, which I am rather surprised was not foreseen by its founders. This lay in the method of selection of Associates, which depended on the writing of a 10,000 word dissertation on a subject with which the applicant was connected in some way. But that almost certainly ruled out active members of the practising farmer section of the industry. It virtually confined applicants to those engaged in education, research or the commercial sector, or to older people with time on their hands. Those engaged in research though were not likely to apply, since their work would be published in scientific journals. There was also a risk that some people would simply use it as a means of furthering their career prospects, which was far from the original intention for the scheme. As a result few applications were received, and virtually none from practical farmers. The Council made some direct elections to the Fellowship each year, to keep the show going. By the time that I was elected as a Fellows' representative on the Council in 1979, it was quite clear that some radical changes would have to be made if it was to survive. An old Cambridge contemporary of mine, Harold Hudson, a prominent Norfolk farmer, was Chairman of the Council at the time. He made some very sensible suggestions for re-writing the rules to make it more attractive for practical farmers. Essentially he recommended removing the necessity for the 10,000 word dissertation, with admission to be primarily dependent on a very searching interview by two experienced assessors appointed by the Council. This was not acceptable to the official RASE representatives, who insisted that the Society should set up a working party of its own to review the scheme. Its report, which took over a year to produce, was quite unacceptable to the full Council, and after further discussions the RASE withdrew unilaterally from the scheme. But the Royal Welsh, Royal Highland and Royal Ulster Societies, with the support of the English Fellows' representatives, decided to set up a new Council under the chairmanship of Meurig Rees, CBE, to reorganise the scheme.

The English Fellows felt that it was essential that England should be involved if possible, and, after consultation with the membership, an independent committee was set up, which was awarded two seats on the new Council. I was elected to chair this committee and with Ian Gibb, from Reading University, was appointed to the Council, which then set about the task of creating a scheme which could meet the objectives laid down in 1970.

The Cotswold Community

A commitment of quite a different kind arose in 1974, though it had only a fairly tenuous connection with agriculture, and certainly none with agricultural developments. But it has turned out to be a fascinating and rewarding experience to have become involved in it — and remains so up to the present time. It arose from a letter from a complete stranger, Richard Balbernie, writing from a body called the Cotswold Community, near Ashton Keynes, just south of Cirencester. I was vaguely aware that there had been an approved school there from hearing broadcasts given by the previous head of the institution. Richard Balbernie informed me that the school had been closed down. It was now owned by a trust and it had become a therapeutic community for the rehabilitation of seriously maladjusted boys, who kept on coming before the courts for various reasons. He hoped that I might be persuaded to go over to advise on what they might do with their 300 acre farm.

The school was one of the first of a small number of centres for maladjusted children, set up to explore the principles developed by the late Donald Winnicott, and further exploited by Barbara Docker Drysdale and Richard himself, among others. Treatment in a therapeutic community is in many ways the exact antithesis of that followed in most young offenders' institutions. In the latter strict discipline and spartan regimes are supposed to cure young delinquents, even though the statistics show that a very high proportion of such offenders come back before the courts again and again. Essentially the treatment in a therapeutic community depends on trying to get back to the root cause of the disruptive behaviour. By sympathetic understanding and psycho-therapeutic methods the children are helped to re-establish more trusting relationships with both the adult world, and with fellow members in the community itself. Many of the case histories of those referred are acutely disturbing, dating back to very early childhood, and often involving broken homes and lack of parental love. It may take several years to re-establish normal human relationships and self-esteem. The dedication of those who care for these boys and almost nurse them back to normal living and more settled and ordered lives is quite beyond praise. The success rate is high, as is the cost, because so much individual attention is given for several years. But at least the state is saved the cost of probable prison housing in the future.

The estate had started life in the '30s as one of the German Bruderhof Communities, dispersed after the war. This was a closed self-contained community in which everything was idealistically shared between the members in an almost back-to-nature existence. They had grown their own food on the farm, spun their own clothes, etc. They had put up quite a few buildings, converting others to a meeting hall, and had even created a small burial ground in the centre of the farm. Then came the war, and as Germans they were interned for the duration. After the war, the members had scattered to other Communities around the world, and the estate was taken over by the Home Office as an approved school, which was closed in 1970.

It turned out that the estate was more or less slap in the middle of what was then being developed as the Cotswold Water Park, where huge areas of land sterilised by gravel extraction were being converted into numerous lakes, to be devoted to all forms of aquatic activity. Gravel extraction was still going on in the immediate vicinity and some 120 acres of the farm lay on gravel beds of varying depth, including the community buildings and playing field. The rest of the farm was on higher ground on two sides of a hill where the soils were very mixed, but with quite a lot of silt particles and spring lines on each side of the hill. The farm had very clearly been overcropped with cereals, the soil structure was extremely poor on the siltier land, and very shallow indeed on some of the gravel land close to the buildings.

Richard laid down certain conditions about management, which had to avoid any very intensive activity, such as milk lorries coming in every day, or big and noisy tractors and machinery, which could disturb the boys. The whole ethos should be one of peaceful activity and orderliness so as to create an environment in which the boys' chaotic lives and impulses could be allowed to settle down and adjust to a measured pace of life. At the same time, it would be advantageous if there were activities on the farm in which those boys who might be interested could play some part. As there was a large lake nearby, it would have been possible to put in a dairy unit with summer irrigation, but it would have been expensive in capital outlay. Little money was available, and anyhow, a dairy unit was ruled out by Richard's request for peace and quiet. He had turned down the idea of a pig unit for various reasons, though there had been a small one there in the approved school days, and I expect that the Bruderhof Community had enjoyed some good home made German sausage in pre-war days!

With all these restrictions, there was really not very much else to recommend than putting down some leys, in order to try to restore some soil structure on the worst of the siltier fields, and to get in some sheep and possibly beef cattle. On balance, I tended to favour a sheep flock rather than a suckler herd, as sheep prices were just starting to pick up by then, and sheep and lambs would present more opportunities for boy involvement than would a suckler herd. The alternative to the latter would be a calf rearing unit, buying in week-old calves, and taking them on to mature beef or selling as stores. I was concerned about likely standards of husbandry with very young calves, as things can go wrong very quickly if hygiene is not 100 per cent in such units. I worked out some figures, and it looked as if the farm could just about break even with a two-man labour force (necessary because of boy supervision). It would need two-thirds of the arable in cereals and about one third in leys and a sheep flock of about

150 ewes, and a few beef cattle. I sent Richard my report, which he acknowledged with much thanks.

I heard no more for, I suppose, about three years, when he got in touch again, to say that Wiltshire County Council had now bought the Community from its previous trust owners, and he wanted to set up a small farm advisory group. This would advise the board of managers which consisted largely of representatives of the three political parties on the county council, with, in addition, a number of specialists from the consultants who helped in the treatment of the boys. The chairman of the managers, and also of the farm committee, was none other than my very old friend and post-war student Jack Ainslie, who was farming in partnership with another friend, Alec Gale, near Marlborough. Jack was currently the chairman of the County Education Committee, so he was a very good choice for the job. Another acquaintance from my Association of Agriculture connection, David Randolph, who also farmed in Wiltshire, was a member, too. So I agreed to join the farm group with alacrity.

To begin with, I was just a member of this group, but after a few years, I was made a full member of the management committee, and became much more actively involved in the Community itself, and in its battles for survival that were to come.

The farm group decided on much the same policy as the one that I had recommended before, but to carry rather more sheep. The old manager then retired, and we were able to jack up the cereal yields, in a good year at any rate. But the thin gravel soil over a considerable part of the farm makes it very vulnerable in a dry year, and especially if it is dry in May and June. The advent of oilseed rape and linseed have provided reasonably good break crops in recent years, and with the leys and sheep, and some much needed drainage, soil condition is slowly improving. But with rent and management charges to pay back to the county council, and higher fixed costs than would be found on most commercial farms, it is a difficult job to show a profit in most years.

I was also keen to join the farm group because I had become very interested in the work that Richard Balbernie and his staff were doing. He was a unique character in many ways, with a distinguished war record that one would not have suspected, since at times he seemed to present an almost unworldly approach. This was combined with an acute antipathy to red tape and bureaucracy of any kind. Like many deep thinkers and idealists, he was not always very successful in putting his ideas into terms that the layman could easily understand. He certainly did not take kindly to the rather formal procedures imposed on him by having to conform to county council rules and regulations. He abominated having to waste time, as he saw it, in unnecessary administration — time which he felt could be much better employed in ministering to the boys in his care, to which cause he was completely dedicated. Relations with the new owners could at times be somewhat strained. But with a sympathetic Director of Social Services, under whose department the Community was placed, and with Jack Ainslie to smooth the wheels at County Hall, the Community developed along the lines he wanted.

Richard saw the farm as having three roles to play in the life of the Community. Firstly, it provided an opportunity for a number of the boys to take part in a constructive activity, which could help in their rehabilitation, and to build up a personal relationship with a member of the farm staff whom they could trust. This is a most important part of the treatment for those who have become alienated from society and the adult world. Secondly, the farm provided a settled and peaceful environment surrounding the Community, shutting it off from the stresses of the outside world, and the influences likely to increase tension in the minds of disturbed boys, especially recent arrivals. Thirdly, in a purely practical way, it provided something of a *cordon sanitaire* round the Community, in the event of a boy taking it into his head to decamp and make for home. These advantages still very much apply today, and will remain so long as the Community continues to exist on the site at Ashton Keynes, which has become a matter for speculation in the past few years.

COLLEGE RESPONSIBILITIES

From the early days, I had always looked after the interests of undergraduates on our degree course from Christ Church, St Edmund Hall, and Merton, none of which colleges had Fellowships in agricultural science. I organised tutorials for them each term, saw that they were attending lectures and classes, set them end of term tests and maintained a liaison with their college tutors. In 1974 Robin Offord, a brilliant biochemist in Christ Church, who looked after all its biological undergraduates, decided that he wanted to reduce his workload. He also felt that the college, as one of the largest landowning colleges in the university, ought to have a Fellow with a knowledge of agriculture. So he asked me whether, if he could pull the right strings, I might be willing to accept a Fellowship.

I was very much in two minds about it. Firstly, because I wasn't sure that I could find the time to do the job properly, and take on the additional responsibilities of a College Fellow. These involved attending governing body meetings, academic committees, generally entering into the life of the college through lunching and dining there and establishing contacts. The second reason was that I had only some six more years to go before retirement, and wondered whether it would all really be worthwhile for such a short period. But against such arguments there was the fact that it would be very advantageous to the Department to have another member of staff with a full college fellowship, since this would help to ensure that college places could be found for worthy applicants. I would be allocated a number of places for biology candidates, and hopefully some of these could be filled with those wanting to read for our degree. Furthermore, if I did not blot my copybook, there would be a possibility that when I came to retire, the college might be prepared to offer a Fellowship to my successor. But, of course, the overriding consideration was that a Fellowship at Christ Church is widely regarded as being highly prestigious in the academic world, and in the outside world too, I suppose – something that most people would sell their souls to obtain - and it would appear to be completely crazy to turn down the opportunity. So I told Robin that I would certainly take it on if it was offered, and would just have to find the time to fit in the extra work.

It took a bit of time for him to work the oracle, but in due course I was invited to dine in Hall with the Dean one night (the Dean of Christ Church is the Head of the College) and to meet a few of the Fellows for sherry beforehand. This was quite obviously to be given the 'once-over', and to make sure that I looked reasonably presentable, and was not given to eating peas with a knife. The Dean at that time was a most charming clerical academic of the old school, who reminded me in a way of my dear mentor of wartime days, Tom Sutton, in his manner and courtesy. He left us soon afterwards to take up one of the ecclesiastical chairs at Cambridge, and then was elected as Head of one of the colleges, thus achieving what must be a unique 'double' for modern times, of having been elected as Head of a college at both Oxford and Cambridge. His elder brother, also a distinguished clerical academic and Head of another Cambridge college happened to have been about a year behind me at school. We both got our third fifteen school rugger colours in the same year, so this broke the ice at dinner! I must have passed the appropriate tests, for not long afterwards I was offered an official Studentship (which is what Christ Church call their Fellowships) in Biological Sciences. I had responsibility for the initial selection of candidates each year for reading either zoology, botany or agricultural science, and their supervision thereafter.

Christ Church, as those who are familiar with Oxford will know, is physically much the largest of the colleges. It has the huge Tom Quad, a large chapel which serves as the cathedral for Oxford, the Great Hall, with its impressive staircase and unique array of portraits, where Charles I held his parliament during the Civil War and a magnificent library. In addition, it houses a unique tower built by Sir Christopher Wren in 1681, over the main entrance. The tower's clock, Great Tom, tolls at five past nine each night in memory of the 101 original students of the college. I reckon that I am a down to earth and normally unresponsive character, but I have to confess that I never went into college without feeling some sense of awe, and indeed of pride in belonging, even in a very humble way, to such an historic institution. The past is all around one, in spite of the roar of traffic, and the hordes of visitors thronging Tom Quad. In fact, the tourist problem became so acute some years ago, with inquisitive visitors entering college rooms, that the main entrance had to be closed to visitors, and an entrance charge imposed. In spite of that, I read that last year some 300,000 visitors passed through the turnstile. I wonder what Cardinal Wolsey, who founded the college in 1525, and Henry VIII, who refounded it in 1545 when Wolsey fell from grace, would think about that!

The 'House' as it is called in Oxford, achieved a reputation at one time as being a college populated by wealthy, and often aristocratic, undergraduates who spent their time in drinking or in riotous living – the sort of Oxford unfairly portrayed in Evelyn Waugh's *Brideshead Revisited*. Before the war there was probably some element of truth in this, though even then, it would have been only a small minority of the members of the college. With Balliol College, it boasts far more Prime Ministers than any other College, so it cannot have been too bad! Today, it is just as cosmopolitan as any other Oxford college, and has an equivalent intake from state schools. One of the first votes that I took part in on the governing body was whether we should admit women, which was passed by a very substantial majority.

My new colleagues turned out to be a most welcoming and hospitable group of men (it was to be another two years before the first woman was elected to a Fellowship) and any concern that I might have had about fitting in to such august company was quickly dispelled. Of course, the Fellowship body of any college, or common room for that matter, will always contain a very wide cross-section of types and per-sonalities, and we were no exception. We ranged from the ivory tower academic to the pragmatic scientist, and the abstract mathematician to the practical geographer, or even more earth-bound mortals, like myself. We had the additional advantage of having the cathedral canons in the common room, which added another pleasant dimension to the society. Some of one's colleagues were easier to talk to than others, but lunching or dining in college was always a pleasurable occasion, since one never knew what subject one might be expected to show an interest in next. It was a unique experience, and I must thank Robin Offord for having worked the oracle in getting me elected. The closer contact with my own undergraduates in college was also for the most part highly rewarding, even though the added responsibility for their welfare could prove slightly taxing at times. For example, when one of them failed to turn up for the end of term interview with the Dean, and had to be fished out of the nearest pub to present himself – better late than never!

So the decade of the '70s drew to its close, with the agricultural industry in a very uncertain state, fearing what might be in store for it in respect of lower profitability, and greater restriction. Simultaneously I reluctantly faced the prospect of having to retire from what was still a very active involvement in several different spheres of agriculture.

CHAPTER SEVEN

The Closing Years

———❧———

THE start of the final decade and a half of this 60 years'
review should have witnessed my retirement on 30 September 1980, my 67th birthday. But Oxford is full of little
precedents established over the centuries, and one of these is
that anyone born within two weeks of the start of the
academic year is entitled to stay on for another three terms if
they wish. The relevant date that year was 1 October, so my
parents must have been exceptionally prescient in my conception to ensure that I was born the day before. It meant
continuing till 31 July 1981, which was good as I was not at
all anxious to go before it was necessary. It did mean, though,
delaying some improvements on the farm which had already
been deferred for a couple of years, so as not to jeopardise the
freedom of my successor to do what he wanted.

But it was just as well that I had to go then, for what I had
feared for the future of our Department finally came about in
1981. The time had come round again for the University
Grants Committee to review the agricultural departments in
the universities, and it had already leaked out that some
courses might be abolished. That might have been fair,
if there really were too many of them. But at the same
time as the UGC was saying that there *were* too many,
the Council for National Academic Awards (CNAA), the
controlling body for awards at polytechnics and equivalent
institutions, was granting degree status to colleges which
were not universities at all. In agriculture this meant that
Seale Hayne, Harper Adams, and the Royal at Cirencester
would attain degree status for their courses. Either the left

hand did not know what the right hand was doing, or the UGC, pressurised by the Government to save money, decided to close its eyes to what was going on in the colleges. As far as Oxford was concerned, other influences may have been at work. Suffice it to say that the Oxford course which was chopped, even though it was not in straight agriculture but in agriculture and forest science, was unique in Britain, was recruiting well and fulfilling a very useful purpose.

What really saddened me was that nobody in the Department seemed at all willing to put up a fight about it. David Smith, our new professor, rightly or wrongly took the view that there was no point in trying to fight the decision, since it was a fait accompli, and that the only way for the Department to survive was for it to combine with the botany and zoology departments in a new umbrella Department of Biology.

There had always been a very strong plant science lobby in the Department, and animal, soil and economics interests had had to fight hard to maintain their corners, and restrain an excessive emphasis on plants. But we had previously resisted this pressure to specialise. I felt bitter at the time that I had no inkling of what was being done until it was all tied up. I suppose that was fair enough, as I was retiring and would therefore have nothing to do with the course in the future, and I might have been expected to rock the boat if I had known about it sooner. As it was, all I could do was to write a letter to *The Times*, in the usual way, about the absurdity of the UGC saying one thing and the CNAA another, and the iniquity of killing off a popular and worthy course. I got quite a lot of support from former students and others, but it was too late.

If I am honest, my indignation was really due to the fact that agriculture had finally lost the battle with biology, and that the small, friendly, well-integrated department that we had fostered for the past 35 years was to be no more. I was quite sure that it would become part of a large amorphous conglomerate in which the remaining agricultural interests would be steadily whittled away. To be fair to those who effected the change, the present course does seem to be a popular one which recruits in very considerable numbers, and some of its members two years ago actually initiated a Young Farmers' Club, so agriculture still survives, though the students' Plough Club has died. The farm has been retained too, mainly as a commercial unit, which provides a range of facilities for biological research of various kinds.

But, if there was a feeling of great sadness about the demise of the course, this was more than cushioned by the extraordinary parting gift that I received when the time came to go. One or two of my colleagues decided that there was not much point in giving me a clock, or even a computer (I think they knew that I'd never be able to master it!) but they thought that I'd appreciate it if any money they collected might be used for the future undergraduates. They set up a committee with Chris Congleton as chairman, and circulated old students, Oxford Conference chairmen, other people that I knew in the farming world and anyone else they could think of. To my utter disbelief they collected a sum which, with covenants, amounted to some £20,000. This was converted into an official university trust fund, to be devoted to making annual travel grants to undergraduates who wanted to go abroad in vacations, to widen their knowledge of agriculture and forestry. The donations included some which were extremely generous, though it would be invidious to name the donors here. I found it very difficult to express my feelings when they gave me the news, and I still do so today. I would like once again simply to say 'thank you' to those who were so generous. In the last 13 years, well over 100 undergraduates have benefited from grants from the fund, which now provides some £2,500 a year for distribution.

In a way, I suppose that I have achieved some sort of immortality in that the 'Mike Soper Bursary Fund' is recorded as an official university fund. But I am not so naive as to think that this means anything at all, since no undergraduate applying for a grant now has the slightest idea who Mike Soper is or was! I hope though that the recipients will be grateful to the original donors, whose names are beautifully inscribed in a book presented to me on retirement.

I am convinced that the great secret about retirement

is not to retire at all, but to ensure that when the day arrives there is enough on the plate to keep one comfortably occupied, and I don't mean doing all those things about the house and garden that have been put off for years. The more active life has been before, the more important it is to maintain activities at a reasonably high pitch. The Soper theory is that it is essential to keep the blood running to the head, which entails plenty of brainwork and not just golf and gardening. I had seen far too many people go rapidly downhill both physically and mentally from lack of work to want to go down that road. So when my day came, I still had three major jobs, and was a member of a number of other smaller committees. I was still acting as Chief Assessor to the NCA Board, with a steadily increasing work load as more and more new college courses on subjects ancillary to agriculture came on stream. The only difference now was that I no longer had my treasured secretary, Gillian Bendle, to type the odd important letter, but had to do all the office work myself at home, and learn a few secretarial skills into the bargain.

Apart from the NCA Board, I was deeply involved in writing for the Association of Agriculture, producing a new series of booklets. Within two years, Professor Michael Wise retired from chairing the Executive Committee, and Joan Bostock and members of that committee put great pressure on me to take it on, which brought extra duties and more trips to London. I was still chairing the Compton Farm Committee, was getting more involved with the Cotswold Community, and within two years was chairing the English Panel of the Fellowship Scheme of the Royal Agricultural Societies (FRAgS), so there was plenty of opportunity to keep the grey cells active.

The strangest thing was that I had expected to miss quite seriously the contacts with students and teaching, and with the day to day control of the farm. Yet when the time came, I hardly felt any sense of loss at all. I suppose that this was because there *was* so much going on elsewhere, I hadn't got time to think about it. Added to that was the realisation that my little cottage was far too small to provide both office and living space, when occupied full-time, so I had to look around for an alternative. At last, after eight months on the market, and just when I had decided that I would never be able to sell it, an offer arrived, which I rapidly accepted. I was then extremely fortunate to find just what I wanted, namely, a three-bedroomed bungalow with a nice garden in a very pleasant village near Wallingford – one of the few really unspoiled towns left in the south of England. So clearing up the junk of 30 years and moving took up time as well.

What I did find on retirement was that in spite of having the Compton and Cotswold farm committees, I quite quickly began to get out of touch with practical farming matters. At first, it was just small technical details such as cereal varieties or new herbicides, but after a few years it was more important things like Ministry rules and regulations, subsidy schemes, or European legislation. Accordingly, this final instalment on agricultural change must perforce be rather secondhand, which is inevitable as one becomes more of an observer from the touchline than a participant on the field of play.

FARMING PROBLEMS IN THE '80s

I am sure that most farmers would agree that many of the doubts and fears about the future that had begun to build up in the late '70s, came to be realised in the next decade – and quite soon at that. The warnings about overproduction became reality with the first of the quota restrictions placed on dairy farmers in 1983, closely followed by co-responsibility levies on cereals.

Then when these proved inadequate to curb production, came the much reviled Set-Aside Scheme in 1988, initially on a voluntary basis. But this was not effective either, for much of the land that was taken out of production for the next five years was of pretty poor quality, so that the overall reduction in yield was quite small. It was to some extent offset by the participating farmers who put more into their remaining better land in order to try to make up for the lost yield from the set-aside land. It was much the same story as the previous efforts to reduce milk production under the Mansholt Plan, when farmers were paid to get rid of dairy cows, and culled all their old and lowest yielders, but fed their better ones a bit more to make up for the lost milk. It was mainly the failure of that plan that led finally to milk quotas.

One of the immediate effects of milk quotas was to concentrate farmers' minds on more efficient management. If they were not allowed to produce more (or were heavily penalised if they did so), could they produce it more cheaply by reducing input costs? There was not much that they could do about most of their fixed costs, such as rent and direct labour, since the number of cows could not be increased to spread those costs. But there was still considerable scope on many farms to reduce variable costs, especially in purchased feeds. This meant primarily a better use of grassland and the making of better quality silage. It was a lesson that some of us had learned many years before – that the more cake that you let the cow eat, the less grass she will eat. That statement is, of course, an over-simplification but there is a strong element of truth in it. The introduction of quotas also stimulated new thinking about techniques such as flat rate feeding, complete diets and the inclusion of a wider range of feeds in an attempt to cut input costs. Quotas certainly jolted a good many farmers out of the high concentrate feeding rut that they had got into.

Another significant change since the early '80s, partly related to quotas and reducing fixed costs, has been in the use of contractors, especially for silage making. It is not worth a small or even medium-sized farmer today buying a new forage harvester at current prices, and especially not when his labour force has been cut to the point where it is stretched to the limit, and when contractors are available to do the job in a fraction of the time with modern tackle. It is quite likely, too, though I have no figures to prove it, that the quality of the silage made by a contractor is, on average, higher than it would be if it was made by a small farm team. The reason is that the job is done so quickly, with the grass all cut at the same stage of growth and put straight to bed in the clamp and battened down, instead of the protracted process that used to be spread over a couple of weeks or more, entailing clamp losses and deteriorating grass quality. The same arguments are true for an increasing number of other jobs on the farm, where the decline in the employed labour force is offset by the use of the contractor with his specialised machinery, and this is a trend which will surely grow in the future, as more and more labour leaves the farm. The latest figures showing a drop of some 60,000 in the total labour force on farms between 1983 and 1993 would certainly support that assumption. It is interesting to note that some 26,000 of that loss was in farmers and their partners, which illustrates the rate at which holdings are disappearing, and why farm size is steadily increasing.

A considerable number of farmers were hard done by when quotas were suddenly introduced, especially recent entrants, though the regional panels did their best to sort out the worst of the anomalies. But it wasn't to be long before a thriving market had built up in buying and selling quota, and then in leasing it, as farmers decided to go out of milk production and cash in on quota value. As the years have gone on, more and more abuses seem to have crept in, to the point where it seems that people with no connection with farming are invading the market as pure (or impure!) speculators. The victims, as usual, are smaller farmers, who simply cannot afford to buy or lease at current prices, whereas the big operators probably can do so. It is especially hard on those who are trying to get a foothold in farming, the sort of enthusiastic youngster who in the past could perhaps get a small farm or a smallholding, and by working all hours without holidays for five years, gradually accumulate enough capital to expand. Today, the quota on such a holding would not be large enough to enable him to do this unless he could supplement it by leasing or purchase. Even that is now generally beyond his means – so another of the traditional ways for young people to get established is becoming blocked. When quotas were first introduced it was impossible to foresee the complications that were to arise from them, especially in regard to land values, tenant rights, new entrants, and so on. One can see no end to them now, so the outlook for small milk producers cannot be good.

While on the subject of milk quotas, it does seem strange to see the amount of money and effort which is being put into the breeding of extremely high yielding cows from primarily American and Canadian Holstein bulls, with all the attendant costs of embryo transplants, high energy diets, three times a day milking, etc. What is the point of it all, when there is too much milk about already? One would have thought that it would be more valuable to be selecting cows genetically capable of giving economic yields of milk from lower quality, cheaper foods. It is possible that a high yielding cow selected under the first system would also be an economic converter under the second, but I don't think it follows, by any means. Thank goodness the dairy industry has had the good sense so far to throw out the milk stimulating hormone, BST (bovine somatotrophin), since there is absolutely no justification for it under present conditions. Quite apart from the humanitarian aspect of frequent injections and of pushing metabolic processes beyond their natural limits, there is the public reaction to be considered, which would assuredly by extremely hostile. It all takes me back to 1937 when the possible feeding of iodinated casein to cows, to stimulate the thyroid so as to increase metabolic rate, hit the headlines. Fortunately, it failed through the difficulty of regulating the dose for individual cows, and because there were a number of fatalities in the trials. I am ethically completely opposed to the use of stimulants of this kind to boost performance in farm animals. If we ban athletes from taking such drugs, what possible justification have we for giving the same sort of treatment to the animals in our care?

As if all these upheavals in the dairy world were not enough to be going on with, there has been the added worry of the sudden appearance of bovine spongiform encephalopathy (BSE or mad cow disease) apparently brought about by the feeding of insufficiently sterilised meat and bonemeal. Both fishmeal and meat and bonemeal have been included in bovine rations for probably 100 years or more with few ill effects – in fact, with very beneficial effects, due to the high biological value of the proteins that they contain for the nutrition of the rumen microflora. Admittedly there were occasional cases of anthrax caused by that organism escaping complete sterilisation – we had a case of this on our own farm in the '70s – but they were not frequent enough to merit exclusion from the feeds. The causative agent in BSE is supposedly the organism responsible for the sheep disease scrapie, sheep brains having been one component of the meat and bonemeal. With a change in the method of sterilisation for a cheaper one at a lower temperature the disease is assumed to have survived and then jumped the species boundary from sheep to bovines. The worry is that having made that jump, it might make a further one from beef to human beings. However, there is absolutely no evidence that it could or has done so, in spite of its similarity to Kreutzfeld Jacob disease in humans, present long before anything was heard of BSE. Furthermore, if there was such a risk, it should surely have surfaced long ago from scrapie which has been endemic in some sheep breeds for at least 100 years or more. The latest reports on the controls put in place as soon as the seriousness of the BSE situation became apparent are encouraging, and one hopes that the disease will be eradicated within a few years, to allow dairy farmers to sleep more peacefully in their beds.

To those with long memories, the demise of the Milk Marketing Board is a worrying development. I came into agriculture the year after it was set up in 1933. I well remember the stories that we were told about previous conditions in the industry: how small producers were virtually held to ransom by the larger dairy companies, and forced to accept rock bottom prices if they were to find a market for their milk at all. One can only hope that dairy farmers today will join Milk Marque, and that those who elect not to do so will look at the alternative baits that they are offered with their eyes wide open. The great difference today is that there are only 36,000 dairy farmers left in the UK, compared with some 200,000 or more in 1933. Nearly 50 per cent of them have 100 cows or more, so they are for the most part experienced businessmen and should be better able to hold their own in the commercial world than their grandfathers were 60 or 70 years ago. But I do feel sad at seeing the old MMB go, with all its ramifications, for it served the industry well over the years. It provided funds for research and a number of other farming activities, which one cannot see the smaller organisations which have succeeded it being able to do. I suppose that this is just another example of change, and we shall have to wait for at least another 30 years for historians to decide whether it was change for the better or not. What does seem certain is that the number of producers will continue to decline, as farm units increase in size, even though it will not be at the same almost catastrophic rate that we have seen over the past 30 years.

One cannot leave the animal sector of the industry without a final update on the sheep situation, which in some ways has recently been the most dynamic sector of all. The period has seen a dramatic overall increase in the ewe population of some four million in the UK between 1983 and 1993, so that overall numbers are at approximately 20 million. Many of the dairy farmers that quit milk production turned to sheep, while on arable and mixed farms the change to more rotational systems incorporating leys, led to largely increased ewe numbers. Only a small number of farmers opted for beef, which was in over supply, and outdoor pigs were restricted to certain areas. In addition, many existing grassland sheep men increased the size of their flocks in an attempt to maintain incomes, as did hill farmers for the same reason, encouraged by LFA supplementary payments. Nobody driving round the country in recent years can have failed to notice the presence of sheep in increasing numbers. This is particularly the case when there are lambs at foot, at the high stocking densities now being employed; and this applies in almost all areas of the countryside.

In 1992 came the rationalisation in the subsidy system for both cattle and sheep, and the institution of a headage payment for ewes, tied to a quota. This is certain to restrict the expansion in ewe numbers in the future.

All the extra lambs have to be found a market when their short lifespan comes to an end, so the other major feature of the past few years has been the expansion into the export market. At first, this seemed to be centred on carcase exports to areas such as the Middle East. But increasingly through the '80s and early '90s the European market was opened up (probably *one* advantage of being in the EU at any rate) with Spain, France and Italy taking increasing numbers. The difference here was that the trade was in live lambs, and not carcase meat, and this is where the troubles of the winter of 1994/5 began, when the export of live calves and lambs first came under pressure from the animal welfare lobbies.

The most significant change in the cattle sector was the steady increase, too, in the export trade for young calves, required by continental producers for the production of white veal. Insufficient calves are available for this purpose in Holland, France and Germany, where the consumption of that type of veal is very popular, and for which the British consumer has never acquired a taste. Unfortunately the methods used to produce white veal – close confinement in crates, liquid feed for three months, and a restriction on the amount of iron in the diet – are regarded as cruel by the animal welfare lobby and are banned in Britain. Rightly so, I think. The outlet for such calves for the more mature pink veal production (which is legal) only accounts for a tiny proportion of the calves born in the dairy herds. The market for calves for rearing for mature beef is also limited, while relatively few are required as heifer replacements in the dairy herds. Previously true dairy type calves were put into the market in the first week and bought for a few shillings, and slaughtered for the manufacture of by-products, or conversion to meat and bonemeal. But many of today's calves are of much better quality, fathered by beef bulls through AI, and they make good veal. It would be a waste of potentially good meat therefore to kill them at a few days old if exports were stopped and it would deprive farmers of valuable income. Short of persuading continental producers to alter their methods of production, it is difficult to see what the ultimate solution will be. One cannot help feeling that enterprising farmers at home could use a lot of these calves themselves, and build up a market, with the co-operation of the big supermarkets, for a form of British veal. It would need a big advertising campaign, but if the cost was shared between the producers and the retailers, it could prove to be a successful venture, and possibly create an export market for the meat as well.

In the case of live sheep, which often have to travel much further distances in Europe before reaching their point of slaughter, I am convinced that slaughtering should be done at home, and lamb exported in carcase form. But the prospect for that is not improved by the Government allowing the closure of many abattoirs, owing to the imposition of unnecessary EU regulations. One reads – and sees on television – that these are widely ignored elsewhere in Europe. I hope that British farmers will come to see that sending live lambs packed together in transporters for 700–800 miles, in addition to a sea crossing, is not really humane, and that they must co-operate to ensure efficient home killing and the promotion of carcase exports. If they don't, they will ensure the continuation of public hostility, which is certain ultimately to make life a great deal more difficult for all farmers in future.

We do seem to be entering a new phase in animal production where public opinion, often influenced by emotive pictures on television, seems certain to impose further restrictions on the ways in which animals are reared and kept, and one cannot help thinking that the banning of veal crates in 1988, and the forthcoming banning of sow stalls in 1999, may only be precursors of other constraints, which will almost certainly mean higher costs of production.

In much the same way that animal producers have had increasingly to tailor their products to market requirements, for example, much leaner pig and sheep carcases, so now are cereal producers having to do the same if they are to obtain that little bit of extra premium price. This can make all the difference between profit and loss in a market which is normally oversupplied. Certainly far more has been heard about grain quality over the past 15 years than previously. This, of course, was always the case with malting barley, but with the huge expansion in wheat production it is now Hagberg numbers, specific weights and nitrogen content which determine what the market will pay, whether the direct purchaser is a miller, a co-operative or a grain trader.

One of the features again of the recent past is the increased number of grain marketing co-operatives – a principle preached since the turn of the century, but only in the past 20 years or so coming to reality through pressure of market forces. The larger co-ops and the very large grain traders have the signal advantage that they can test and then bulk up relatively small consignments from farms into large lots. These can be sold either to the large millers or compounders, or nowadays for export, in shiploads of uniform quality. It is, indeed, a far cry from 1934, when we imported 6.3 million tons of wheat, and 750,000 tons of barley, of which 40,000 tons came from Russia. It is also a long way from the farmer going off to the local market with a small bag of grain after the threshing machine had been in, and hawking it round the local merchants to see who would give him the best price. It's only about 50 years since practically all grain was traded in this way – surely change has been all for the best here. For the smaller farmer, whose 1960 storage and handling equipment is becoming obsolete or worn out, the co-operative offers great advantages, though naturally he must meet the charges for handling, cleaning and storage if he wants the whole job done for him. But this is probably more economic than having to find a large capital sum for replacement at today's interest rates. The first thing that I pushed for at the Cotswold Community was to join both a grain and also a livestock marketing co-operative and we have had very good service from both of them over the years.

Apart from Set-Aside, with all its permutations, the other significant change in the arable scene has been the increasing interest being shown in using land for non food crops, as a means of reducing cereal surpluses. Again, nothing really new in that, for flax and hemp used to be very widely grown for fibre production, especially for linen in Ireland, and in the last century quite large areas were planted to copse, mostly for sporting purposes. But now the emphasis is being placed more on energy crops such as rape or linseed as a means of reducing pollution – coppice willow and poplar seem to be getting off the ground at last with the building of two small power plants to generate electricity from chippings. To be realistic, such enterprises (assisted by wind generation in the West and North) are not likely to produce sufficient energy to meet more than quite a small proportion of national demand, but if they could remove the necessity for Set-Aside, or take a few hundred thousand acres out of cereal production, they could be well worth the investment needed to set them up. They could also provide some farmers with income not dependent on food crops.

But really the development that is most needed is a comprehensive policy for the production of home-grown timber. With an import bill for some £5–6 billion each year for timber and pulp, it is crazy for a country which can grow trees as well as Britain can, not to devote a much higher proportion of its land, now that it is not needed for food crops, to afforestation. This is especially important at a time when the rain forests are being depleted at an alarming rate. It is quite unbelievably short-sighted in view of the timescale involved in growing trees. The Government claims that it is giving generous assistance to farmers and landowners to plant trees, but in fact the grants just don't seem to be adequate for the purpose. The investment of a few extra million pounds or so now would save many millions in the future, when world reserves are depleted, and prices rise accordingly.

Though farm size has continued to increase, another significant trend has appeared over the past decade or so. This is a large increase in the area farmed under various contractual arrangements by big operators, for other farmers or landowners in their vicinity. This is partly due to the unwillingness of landowners to tie up land in long-term tenancy agreements, but also due to the high cost of machinery and equipment. The time comes on many smaller or medium sized owner-occupied farms when tractors, drills, combines and so on need to be replaced, and the cost of such replacements may be greater than the relatively small acreage can justify. At the same time, there may be a farmer nearby with a big acreage and powerful equipment who is looking

for additional land over which he can spread his equipment costs. A contractual arrangement can suit both partners
admirably. In the past few years, a number of the larger land
agency firms have worked out a basis for the establishment
of suitable contracts which appear to be working well in
practice.

Contractual arrangements are also spreading to other
crops, for example, to potatoes. An arable farmer will let out
fields on an annual contract to someone who is very well
equipped and an expert producer, who will grow and market
the crop through an established market organisation. In this
way, the contractor can obtain a large acreage and tonnage of
crop for marketing, while the owner gets the benefit of an
excellent break crop and a payment which will pay the rent
and leave some profit in addition – an ideal arrangement for
both parties. This is not new either, since it is the kind of
arrangement which has operated for very many years in the
Cotswolds, with growers in the Vale of Evesham taking land
on the high ground to grow Brussels sprouts. Arrangements
of this nature will continue to expand in the future, as
machines get larger and more expensive, and as marketing
organisations demand bigger consignments of a very uniform
product, which can only be produced by very specialist and
experienced growers able to operate on a large scale. The
change in the tenancy laws should favour more flexible
arrangements to be made between landowners and local
farmers for short-term lettings for perhaps specific purposes.

But looking to the future of crop production, one further
question remains to be answered. At what levels of intensity
will such land be farmed in the years ahead, in the light of
two very possible eventualities? These are, firstly, restrictions
which might be imposed on environmental grounds on the
amounts of fertilisers and pesticides that farmers might be
allowed to use, perhaps in defined 'high risk areas'. There is a
possible foretaste of this in the Nitrogen Sensitive Areas and
Nitrate Vulnerable Zones schemes. Secondly, factors such as
the GATT settlement or CAP reform may bring the price of
cereals in the EU down closer to world market levels, while
increases in the cost of fertiliser and pesticide inputs might

make it uneconomic to use them in large amounts. I think that this is likely to happen though the critical point has not yet been reached, at least for those farming on better quality land, where relatively high input farming still appears to give the best return, regardless of possible environmental hazards, such as residues in drainage water.

However, there does seem to be an increasing awareness among progressive farmers that using reduced inputs, but targeting them very carefully to the projected requirements for each specific crop, may provide as good a margin at the end of the day as would be provided by a high input policy. That this is so is proved by the appearance in the past six years of what might be called the 'three L' initiatives, and the support that they are receiving from farmers. These are LIFE, LEAF and LINK, which acronyms for the uninformed, stand for: less intensive farming and the environment (LIFE), and linking environment and farming (LEAF). The LINK project is an ADAS initiative which involves collaboration between five commercial farms on which the yields from crops under a lower input approach (LIA) are compared with those obtained from a current commercial practice (CCF) system, with careful recording of the costs and returns from each alternative.

The LIFE project is based on detailed experiments at Long Ashton Research Station, which are now being replicated on two commercial farms, while the LEAF project is playing an important role in coordinating the experiences of farmers who are adopting reduced input and environment friendly systems. Through a number of demonstration farms results are disseminated to other farmers. I'm gratified by all this in view of a long-standing predilection for a moderate input approach, and dislike of the very high input philosophy both from an economic and from an environmental viewpoint. It was a happy experience to organise a conference in November 1994 for the Fellows and Associates of Royal Agricultural Societies on this very topic, and to produce what one hopes was a significant report on the present situation. The conclusion is that the days of very high input arable farming aimed at maximum output per acre, regardless of environmental consequences, are nearly over. The future must lie with an attempt to target the level of input to each field which will give optimum production of a high quality product – one which will command a premium price in the market. Under very favourable conditions of soil and climate, this may mean quite high levels of inputs according to the perceived needs of the crop in question, but avoiding a blanket covering or prophylactic sprays. Under other conditions, it may mean considerably lower inputs, since the yield potential of the land may be less and the optimum return will be obtained from a restricted regime. The difficulty lies, obviously, in deciding what each crop does actually need. This should get progressively easier as time goes on and research develops more sophisticated methods of sensoring for both nutrient requirements and disease status. In view of the scientific advances that I have seen in my own lifetime, developments of this kind are certainly not beyond the bounds of possibility and may, in fact, be just around the corner.

The financial climate in farming improved considerably in the early '90s due to the introduction of acreage payments, some relatively poor harvests which helped to keep prices up, and to a devaluation of the currency. The only sector which did not benefit to any extent was upland farming. But this must surely be only a temporary relief since lower prices would seem to be inevitable in the context of bringing the EU price regime closer to that of the world market. Though world population seems certain to increase the overall demand for food, this can only be of benefit to European farmers if their prices are low enough to compete on the world market without the expenditure of large sums in export subsidies, which would not only be contrary to GATT, but also unacceptable to EU taxpayers.

The other factor which is going to make farming in Britain more difficult, and in some ways unattractive, is the pressure from the environmental lobbies, whether related to conservation measures, free access to the countryside, or animal welfare issues. A recent article in *The Times* postulated that with the opening up of Eastern Europe and the development of its agriculture along Western lines, there would really be no future for British farmers as food producers. The suggestion was that they will largely exist, and be paid by the state, to keep the countryside in a fit state for the recreation and enjoyment of the rest of the population. This was, of course, about 80 per cent rubbish (and it was only with great restraint that I refrained from rushing to my desk and writing still *another* letter to the editor to say so!); but there was a small element of truth in it. It was rubbish because there are many areas in the country, farmed by the calibre of the present generation of farmers, which should remain competitive in production with anything from Eastern Europe, and with the advantage of having a 58 million population on the doorstep to feed, and a relatively equable climate in which to operate. But the element of truth is that there are also quite considerable areas which are marginal for food production and only viable with the aid of quite high subsidisation. In the future *if* food is plentiful in the EU and the world, it *might* pay the state better to maintain these largely as recreational areas, and to pay the owners or tenants to keep them as such. It is not a particularly pleasant thought, but it might help to relieve environmental pressures on other more productive land, and provide a better living for those who occupy the land. But current opinion seems to be moving towards the view that food will not be so plentiful as to put such land out of production.

There always have been some areas, of course, which *have* gone in and out of production according to the economic state of the industry, and those of us with memories of the '30s or who were responsible for reclaiming derelict land during the last war, could name quite a few which would fall into this category. A prime example can be seen on Dartmoor where old field boundaries are visible quite high up the hill, marking the limits of the 'new takes' which were brought into cultivation in the prosperous times of the mid 19th century. They reverted back to virtual moorland in the Depression, and most were not even thought worth reclaiming in the last war. It goes against the grain for someone who has spent most of a lifetime trying to increase food production to write in these terms, but there is clearly a limit to what can economically be kept in production if adequate food can really be produced much more cheaply elsewhere. What would be highly undesirable would be for such areas to be neglected completely and left to become unpopulated wilderness. If the nation does not need them for food, but can find a recreational use for them, then it should be prepared to pay for maintaining local communities in a healthy condition, and avoiding depopulation over large areas of the countryside.

However that may be, I would still maintain that a high proportion of the land in this country could be farmed as profitably as that in most neighbouring countries in Europe, provided that farmers in those countries are not given hidden assistance denied to farmers here by our Government. The technical and managerial expertise of British farmers is, in general, as high or higher than elsewhere, and we are quite rapidly catching up in marketing organisation where previously some competitors had the edge on us.

But the attitude of the writer of *The Times* article is only too symptomatic of what the agricultural industry has to contend with from a media which seems to be continually hostile and critical, and in which farmers are perpetually portrayed as the villains of the piece. Yet I do not believe that this is really the true feeling of the average man in the street. One has only to attend a few farm open days to realise how deep an interest there is in farming in the minds of large numbers of the public. A lot of goodwill is shown when actual facts and live issues are presented to people in a realistic and noncontroversial manner, in the setting of a working farm. Just how the situation can be changed is a question to which I have no real answer, in spite of 20 years' work with the Association of Agriculture.

THE ASSOCIATION OF AGRICULTURE

Our work at the Association of Agriculture expanded considerably throughout the '70s and '80s. About the middle of this period we were fortunate in obtaining a grant from the

Countryside Commission to coordinate a programme of farm open days throughout the country. As a result of this we were able in due course to publish a booklet each year, listing those farms which were to be open to the public. We gave advice to local organisations when required, and it was good to see how farmers responded, often under the stimulus of local NFU branches. About this time, too, an increasing number of farmers decided, as a form of diversification, to cater for parties of school children, or adult groups, by converting some of their buildings for instructional purposes, and laying trails round the farm. These have done excellent work in familiarising the next generation with farming matters, and are still doing so today. In the early days Joan Bostock was much sought after for advice in the setting up of such farms, and wrote an excellent booklet on the 'do's and dont's' of taking school parties onto farms.

In view of the concern expressed at the present time about the image of farming and the relationship between farmers and the public, it might be worth recounting in slightly more detail the work that the Association actually did over the 20 years between 1970 and 1990, and up to its demise a year or two later, due to lack of financial support. Firstly, there was the work with schools, consisting of the provision of information sheets on topics about which teachers and pupils might require unbiased facts for examinations in either agriculture, biological subjects or geography. Then there were the booklets dealing in much greater detail with specific subjects, such as soils, grassland, dairy farming, cereals, pigs, sheep and so on. In the mid '80s, the Association played a major part in a new publication in the 'Finding out about' series from the Cambridge publishers, Hobsons. This was *Finding out about Farming*, which was a profusely illustrated booklet which had large sales in the schools.

Each year the Association ran three regional teachers' seminars in different parts of the country, attended by some 50–60 teachers from the neighbouring counties. At these, the format that we evolved was to start with a paper on the agricultural industry, given by very authoritative speakers such as Professors Ian Lucas, Gordon Dickson, John Webster, or Tony Harris, then Principal of Harper Adams. This would be followed by one on conservation and the environment, usually by Eric Carter, then the Chief Adviser to the FWAG groups. A paper on land use in the countryside from the NFU's chief adviser, Brian McLaughlin, would follow, and then a final one on the use of farms in teaching given by Jim Tyrell, a very experienced educationist. Plenty of time was allowed for discussion, which was always of a very constructive nature, in which the teachers showed a keen interest in getting straight facts about modern farming problems. The afternoon was spent in group discussions, investigating a number of different aspects of using farms for educational purposes.

At these seminars, staff members were present to deal with a very wide range of questions, and we put up our display boards and sold a lot of our own or other relevant publications. In addition to these seminars, we had a stand each year at the big two-day meeting of the Geography Teachers' Conference, in London, where a lot of enquiries were received, literature sold, and contacts established. I do find it quite maddening to go to meetings now and hear farmers and others say, 'What we ought to be doing to counter adverse propaganda is to work on the younger generation through school teachers', as if this was something new. We were doing it for years. In addition to the seminars the Association put on a conference in London each autumn on a topic of importance, which was attended largely by teachers or interested members of the general public.

During this period the volume of enquiries coming in mainly from schools greatly increased, particularly as teaching methods began to change, and project work became fashionable. It became necessary for the Association to employ extra staff to deal with these. We were lucky to recruit Jack Whelan, an ex-graduate from Wye, as assistant to Joan Bostock. His great energy and enthusiasm brought a new dimension into the work. He wanted to change the name of the Association to that of the Food and Farming Information Service. Although the original name was never very satisfactory, we did not think that Jack's proposal was quite right either, as it put the emphasis rather too much on food. So the change was not made, though it was adopted as a sub-title. Jack took over the responsibility for the advisory service, which could mean answering as many as 20 queries a day. Some of them needed answers in depth, but others could be dealt with speedily by sending copies of our own or other relevant publications.

In the mid '80s Charles Jarvis, our very active President and a strong source of support to me as Chairman, suggested that we needed a rather different sort of publication to deal with the really controversial issues which were beginning to surface in a more acute form. So we designed a new series of leaflets entitled 'Questions of the Day'. Each one consisted of a series of questions relating to the topic, with answers provided, stating the pros and cons in each case. The first one was on the very topical subject at that time of straw burning, entitled *Must Straw be Burnt?* Then there was *Fertilisers and the Environment*, and others on the use of pesticides, animal fats and human health, feed additives, and so on. These were mainly written by experts in the field and edited by ourselves into a uniform format, and they proved extremely popular, especially at conferences when the leaflets seemed to go at an amazing pace. The last two I wrote for the Association were *Biotechnology in Plants* and *Biotechnology in Animals*. These dealt with techniques, possible exploitation for good or ill, and the ethics of genetic manipulation. These leaflets, too, were in much demand for the subject was something of a 'hot potato' – as it still is.

As the pressures on the industry built up in the early '80s, with more and more exaggerated statements being made by the extreme groups, Joan Bostock and I decided that a book was needed which would make a case for modern farming methods in a balanced way, and one which would also include details of what was actually being achieved in relation to the conservation of the environment by the FWAG groups and progressive farmers. We enlisted the help of Eric Carter, who was a very good friend of the Association and a member of our Executive, and as a result he and I collaborated in producing the book *Modern Farming and the Countryside*. This included case studies of practical farm conservation. We published it ourselves with the aid of a grant from the Ernest Cook Trust, and we were fortunate in having it adopted as a set book for a course run by the Open University. This naturally increased sales enormously, and exhausted our initial print run of 2,000 copies quite quickly. It created a certain amount of havoc in our small office as we suddenly became a book distribution centre, which involved the packaging of parcels of anything between 1 and 20 copies a day. It stretched the small staff to the limit on top of their other work, and I spent a good many hours there myself packaging and invoicing, and learning quite a lot about the book trade at the same time. The book very quickly became out of date due to the introduction of milk quotas and other developments, and Eric and I subsequently revised it, rewrote some sections, added new case studies, and were fortunate in persuading Farming Press Books to re-publish it under the title of *Farming and the Countryside* in 1991, and it is still in print.

The original constitution of the Association allowed for a Council made up of representatives from a large number of organisations, and a smaller Executive Committee. But we realised during the '80s that this was cumbersome and expensive to service, as costs continued to rise. The Council merely replicated the work of the Executive, which had expanded very rapidly during my time as Chairman. A working party was set up, which recommended amending the constitution to allow for just one management body, to be called the Board of Management. Joan Bostock then retired at the end of 1989, after 35 years of devoted service to the Association, and I decided to go as well before the new constitution came into effect.

Funding had been difficult for years, but due to the recession it declined more rapidly. Joan's successor had resigned after a very brief spell in office, and then Jack Whelan left to set up his own publishing and PR business. He maintained the Food and Farming Information Service at the request of the new Board. Then Robin Malim, as Chairman, battled on manfully for a time, trying to raise adequate funds to keep the organisation in being, but it was a losing battle, and defeat was finally admitted in 1993 when the Association closed its doors. But then the RASE, to its great credit, stepped in and took over the Information Service, and Catriona Lennox, who had worked with both Joan Bostock and Jack Whelan, stayed on. She, thankfully, is now based at Stoneleigh and doing an excellent job in responding to the numerous queries which still come in from schools and individuals eager for factual information. But, of course, all the other services that we used to provide have gone, and it is really no credit at all to the industry that, at a time when it was coming under increasing criticism from the public, it could not find the money to maintain those services. I would hazard a small bet that within a few years leaders from the different sectors of the industry will call a meeting and say, 'What we really want is an independent organisation supported by farmers, landowners, the commercial interests and educationists which will collect and present unbiased information about agriculture and the countryside, so let's all get together and set one up' – and the whole thing will start all over again from scratch, with all the extra costs involved. I am afraid that it has to be said that farmers have not in the past been very far-sighted when it comes to raising funds even for research, though here the situation is certainly now a great deal better than it was even five years ago.

MORE CHANGE IN EDUCATION

At the time of my retirement, I was still acting as Chief Assessor to the National Certificate Exams Board, which was actively extending the scope of its operations. The Board seemed content that I should stay on for a time, which suited me well, as it was such an absorbingly interesting job. It was a time of rapid change in most of the colleges, since it was becoming apparent that with the decline in the farm labour force there would not be enough customers in the future on straight agricultural courses to justify the maintenance of the existing number of colleges. Some county councils were already beginning to question whether the needs of the industry could not be met by part-time day release courses. But as the numbers employed in agriculture declined, it became more important that those who stayed on should be trained to a high technical standard. This led to more advanced courses which helped to fill the gap. Farmers, too, were looking at other types of activity, such as equine enterprises, fish farming, and so on, into which they could diversify, so as to help to maintain falling incomes. Most colleges were quick to latch onto this trend and put on courses to cater for them, and there was no lack of takers from young people who wanted a life in the countryside, but for whom there was no room in practical farming.

The Board began to receive a large number of requests to set up examination schemes nationally to cater for such courses, and I well remember the early discussions that took place as to whether it had the authority to include these peripheral subjects under the agricultural umbrella. Very sensibly, it decided that it should do so because they were all countryside activities, there was no other body so well equipped to do it, and we already had a very well-proven and adaptable examination structure. There was the additional advantage that the colleges had faith in us as a Board and were very keen that we should do it.

My job was to tailor examination structures to meet the needs of each course, and to ensure that they met the standards set for the National Certificate in our other subjects. This was generally not very difficult, but before the Board would accept a course for examination, it wanted to be sure that adequate facilities were available to provide a thorough coverage of the subject, and also that there were enough potential students to mount it. This meant, in most cases, visiting colleges with the Chairman and another Board member to ensure that a new course was viable, as the last thing we wanted was to be saddled with courses with very small numbers which were not financially worthwhile. Strictly speaking I suppose that this was not our job, but that of the Local Authority, but we needed to be certain ourselves that the course would be of an adequate standard. These developments were all symptomatic of the changes taking place in the countryside in general at the time, and of the diversification of farming to maintain incomes.

At this point the Board was absorbed into NEBAHAI (the National Examinations Board for Agriculture, Horticulture and Allied Industries). But then along came National Vocational Qualifications and other subsequent changes. But by then I had decided that as I was coming up towards 75, it was quite time to stop being directly involved with young people. So I went in 1987, after completing 25 years as Chief Assessor. It was good to look back and see the progress made in that time – from 24 colleges with straight agricultural courses and some 800 candidates and eight assessors, to 38 colleges with 89 separate courses, some 2,000 candidates and no fewer than 28 assessors. This represented a quite extraordinary expansion in agricultural and countryside education. I do look back with great pleasure to the amazingly friendly co-operation that I always received from the college Principals, and the warm welcome that one always received from their staffs when the summer examining visits took place. I appreciate also the support and friendship of my deputy, John Nullis, who took over when I retired. One forgets in the meantime the endless hours spent vetting examination papers, and then reading partially illegible scripts, when what I really wanted to do was watch the tennis at Wimbledon!

DEVELOPMENTS IN THE FELLOWSHIP SCHEME

Any vacuum that might have been left by leaving the Board was already well filled by an ever increasing workload engendered by the progress made with the Fellowship Scheme. After the new council had been set up in 1983, the first thing was to establish a proper constitution, which strangely did not seem to exist before, and a formal agreement to be signed by the three member Societies. In this regard the lawyers of the Royal Highland did a very good job and charity status was established with the commissioners. A small secretariat had to be set up and immediate financial provisions made to ensure that the new council would have sufficient funds to run the scheme. This was achieved by each society making an interest free loan of £2,000, and the introduction of an annual membership fee of £20 a year for Fellows, and £10 a year for Associates. A registration and assessment fee was also required from new Associates. These fees have not had to be increased in the intervening years, so it was good planning in 1984.

The Council was fortunate in that John Wigley, OBE, the very efficient Chief Executive of the Royal Welsh, was just coming up to retirement. He had sat on the former council for some years, and was very familiar with its work, and he was persuaded to accept the secretaryship, working from his home in Wales for a very modest salary. The other very important task was to overhaul the previous scheme in its entirety and rewrite its aims and objectives. We needed, also, to formulate a set of regulations, primarily for the admission of new Associates, and to set out the guidelines to be followed for the assessment of candidates, to ensure the maintenance of high standards. Again it was fortunate that Edward Griffiths, the former ADAS Director for Wales, was also a council member, and he was extremely experienced in administration and committee work. The council appointed him and myself to prepare drafts to cover these different aspects, which we managed to achieve in the course of a hard day's work at Llanelwedd, the Royal Welsh Headquarters. We concentrated particularly on drafting regulations for the admission of Associates, basing this on the recommendations of sponsors and an in depth interview by two assessors appointed by the council. We removed the necessity of submitting a lengthy dissertation. The council adopted our proposals with certain modifications, and set up four national panels to operate the scheme under its aegis. This has worked very well over the years. All this took some time, but by mid 1984 we were more or less back in business again.

For the first three years I dualled the role of chairman and secretary of the English panel while we were getting established. After two years we set up the sort of committee that had proved so effective in my Oxford Conference days. This involved two new members elected by the English Fellows and Associates each year, with a service time of four years. It meant eight elected members on a rotating basis, and a panel of 12, allowing places for a secretary, a past chairman, and two co-options if required. This has proved to be a very satisfactory format. We spent the first year or two trying to encourage suitably qualified people to apply for associateship, and brought in a few distinguished personalities, who had been missed out while the scheme was in the doldrums, by direct election to the Fellowship. It was not always easy to persuade people to apply, as they would ask why the RASE was not involved, but generally the high calibre of the existing members convinced them that they should become members.

Progress was slow and in order to prime the pump further the council agreed that we could put up quite a few prominent agriculturists for direct election in 1986, and this had the desired effect. That year we held the first spring meeting for the membership, and I stood down as panel chairman, handing over to Robert Bruce, who was later to be elected as Chairman of the Council. The next year we held the first Royal Show reception, with the aid of Bob Bruce and the Midland Bank, and this has become the most popular event of the year, and one to which Fellows and Associates from the other three countries are welcomed.

Right from the start I had been determined that the Fellowship should not just be in the nature of an admiration society, but that it should be made to play a positive role in agricultural affairs. Firstly because it contained within the membership the leading practitioners in the different sectors of the industry, and therefore had a tremendous fund of knowledge and experience available which should be made use of. Secondly, it was a completely independent and non-political group of professional people, who in our context would not have to answer to pressure groups. This concept led to a third annual event, the November seminar or conference in London, at which an important topic of current interest is discussed in depth, and a report is issued. This is widely distributed within the industry and to influential bodies, Government departments and so on. The topics have varied from research, land use in the countryside, to farming within the EU, and reduced input farming. As a result of these developments, the council is now sometimes consulted on important national issues, which gives it a further opportunity for getting its opinions known in influential circles. We must ensure that this function expands in the future, providing an independent semi-professional voice for the benefit of agriculture as a whole.

In 1994, we instituted a fourth meeting for English members in the summer, in the shape of a farm visit in a part of the country distant from the location of the two-day spring meeting. These events are well attended, and are welcomed by members as opportunities for getting together and discussing important aspects of the industry.

The work of the Council has continued to evolve over the years, though it received a severe shock with the sudden death from heart failure of John Wigley in 1989. But we were then very fortunate in persuading J.A.C. (Ian) Gibb, OBE, to take his place. Ian had recently retired as Head of Agricultural Engineering at Reading University, and was one of the foundation Fellows appointed in 1970. He was a strong supporter of the new scheme. He was experienced both in the working of the scheme, and also from holding office in a number of professional bodies, in the management of this kind of organisation. He has proved an immensely efficient secretary to the Council, guiding it with great skill through a continuing evolutionary phase. Another source of strength has been Professor Jim Hall, CBE, formerly of the West of Scotland College and Glasgow University, who has acted as Chairman of the moderator's panel. This is responsible for the screening of all applications for Associateship, the appointment of assessors, and the general maintenance of high standards for admission.

The improved standing of the Fellowship resulted in the very welcome return of the RASE to the scheme in 1991, and this has greatly assisted the work of the English Panel.

The strength of the UK membership is now over 600,

which surely justifies the confidence shown by the Royal Highland, the Royal Welsh and the Royal Ulster Societies in the crisis of 1983. May the Fellowship continue to grow in size and importance as time moves on. Considering the ability and status of the membership, I have every confidence that it will do so, as the old ones amongst us give way to youth (or relative youth, anyhow!).

PROBLEMS AT THE COTSWOLD COMMUNITY

The vacuum created by giving up the NCA was also filled to some extent by becoming a full member of the Managers Committee of the Cotswold Community. Here, serious problems blew up in the mid-80s due to the need for Wiltshire County Council to raise additional money. This was a time when the Government was at the height of its policy of encouraging everybody to stand on their own feet by exploiting every conceivable asset. The asset in this case was the gravel lying under the farm. From this time onwards the future of the Community became linked in with local politics. We had to agree with much reluctance that we could possibly spare a 30 acre block at the furthest end of the farm, well away from the boys, for extraction, even though this would make the farm an even less economic unit than it was already. But we said quite categorically at the time that this must be the limit, since anything nearer would cause serious disturbance to the working of the Community and its remedial function with the boys. There was then a delay, since part of the land lay in Gloucestershire and contained archaeological remains of a settlement. The importance of the site had to be assessed before any disturbance of the land could be permitted.

After the next local elections and a change in the political balance of the council, came a demand for another 70 acres for extraction. This would take out over a third of the farm land, but, much worse, it would bring the gravel workings to within 150 yards of the Community buildings. That this should even have been contemplated at all when the council had originally bought the Community at the instigation of the Home Office with the objective of safeguarding its future, was alarming, to put it mildly. It would mean that the Community might have to close its doors; firstly, because it would be quite impossible to work with the boys under the disturbance, noise, dust, and lorry movements and secondly because of the risk of the boys being attracted like a magnet, either to the water in the pits or to the large amount of machinery on the site, which could be very dangerous. The proponents of the scheme claimed that this was not important as it could all be shut out by boy-proof fences, which of course would be the last thing that the more disturbed boys needed!

So it was decided that the only thing to do would be to relocate the whole Community, lock, stock and barrel, to a new site. This turned out to be a godsend of an idea, since it delayed the action for at least two years, while a search was made across a wide area for a suitable site. Every potential location had to be turned down for one reason or another, and then when the figures for the new buildings, staff houses (essential for this type of institution), site purchase, and 101 other things that would be required, were properly costed out, it turned out that the final cost would probably exceed the value of the gravel available for extraction. So after a huge waste of staff time, that idea was put into cold storage, where it should have been all the time, and would have been, but for the obduracy of certain members.

The next stage of the saga was that a new line was drawn rather further from the buildings, and, though many council members objected, this proposal was carried in the county council. There the situation rests for the moment with the extraction proposal before the planning committee. It leaves a sword of Damocles hanging over the Community, and the hope is that planning will be refused. It has become obvious that a body such as a county council, always at the mercy of political whims which can change at each election, should not own a national resource of this kind. An independent trust has been set up to watch over the interests of the Community, and possibly to organise its leasing from the county council to a national charitable organisation, which would provide more stability in the future. So if anyone reading these words has access to funds, or knows of a suitable trust with such funds, and who would wish to help to preserve a pioneering institution engaged in the restoration of shattered lives, they will be welcomed with open arms. It was a bad day when doctrinaire party politics intruded into local government, and the independent element disappeared.

CHANGE – BENEFICIAL OR HARMFUL?

We cannot close this account of a lifetime in agriculture without a final look at change, and where it has led us. In bare statistics the contrast between farming in 1934 and 1994 is enormous – as great probably as that found in any of the other basic industries, such as coal mining or steel. In 1934 there were well over one million farmers and regular farm workers on the land, but by 1993 this had dropped to some 175,000 bona fide farmers, and 115,000 regular male and female full-time employees. The latter figure is about 45,000 fewer than it had been only ten years before; so the industry is still continuing to shed both farmers and employees at quite an alarming rate. I use the word 'alarming' consciously, because such a loss of population, including families, must be very detrimental to rural communities, and the general fabric of the countryside. The place of the lost workers is largely taken in villages either by commuters, or, more often, by retired people, who, in the nature of things, are less likely to play active roles in local affairs. In some areas, of course, it leads to severe depopulation, and a resultant loss of services, such as schools, buses, village shops and so on. Change, therefore, in this context certainly has not been for the better, and one can only be very concerned as to how much worse things may become in the future.

From the aspect of the individuals that are left, however, change must in general have been beneficial, though one caveat should be entered in the case of the farmer himself. His standard of living, on average, must have improved very considerably, but the average conceals wide extremes – as it did in the '30s also. Superficially, today, most farmers appear prosperous with a good standard of living, but the caveat lies in the increased suicide rate and susceptibility to high levels of stress, which can only be the outward signs of internal pressures and insecurity. Much of this is probably due to the financial squeeze on incomes of the past 15 years, particularly on those with high interest charges to meet. It is probably also to some extent caused by the growing pressure of paperwork and bureaucratic controls, and the increasing isolation of many small farmers, with no employed labour and fewer contacts with the world outside.

The discrepancy between the large scale, business type farmer of the Eastern Counties, with a supporting office staff and access to specialist consultants, and the small scale, isolated, livestock farmer in the West, must surely be greater today than it was 60 years ago. The gulf may become wider still in the years ahead. But I suppose that this discrepancy is no wider in farming than in the world outside, with its immense gap between the £1 million a year business tycoon, and the working man made redundant through no fault of his own with his house being repossessed by the building society, and with nowhere to go – a state of affairs so symptomatic of modern society.

Compared with only 20 years ago, the farmer's lot is not a particularly happy one. He is hemmed in by restrictions on what he is allowed to produce, by quotas, Set-Aside, and continually changing rules and regulations emanating from both Brussels and Whitehall. He is also surrounded by an apparently hostile public, prepared to criticise his methods and disrupt his trading channels, and even subject him to violence in the cause of animal rights. I said to a farmer friend only recently, 'Thank goodness I am out of practical farming now, I just don't think that I could cope with the awful restrictions and the bureaucracy, it would almost drive me mad.' To which he replied, 'Not a bit of it, it's all part of the challenge, and the rewards are still there.' He didn't say, since he is a tactful man, that I was obviously getting old and past it. If his view is widely held, then I don't think that we should be too worried about the future, as people like him will prosper, whatever may be thrown at them. But I am afraid that there are others that will not survive so easily.

But as far as the farm worker is concerned, change can only have been for the better. The days of cripplingly low wages, sub-standard housing, hard drudgery, and working conditions which were always a severe health risk are gone, and a good thing too. Statutory wages, paid holidays, and probably enough money to take the wife to the Costa Brava if she wants that sort of thing, is no more than our workers deserve. They are a body of highly skilled and responsible operatives who can hold their own with technicians in any other industry. I count myself as exceptionally fortunate when running our farm in having had such a loyal and dedicated staff, whose service to us averaged out at 25 years each, at the date of my retirement.

As far as young people are concerned, it will become even more difficult for them to become farmers in their own right in the future, but at least there should be good, well-paid jobs as technicians, with considerable responsibilities. There will also probably be more managerial posts at the top of the employed ladder – and, who knows, they might well be a lot better off, and lead less stressful lives in that capacity, than as small, struggling farmers trying to make a living in an increasingly difficult world. Then, too, in a more science-related industry, the role of the technical or managerial consultant, who has come to the fore so prominently in recent years, could well expand further, creating interesting jobs for those who cannot farm in their own right.

In the machinery and engineering world change has been just as dramatic, though in this case a comparison of simple numerical statistics is meaningless, since there is no comparison between the 1930s Fordson, and its 1994 counterpart. It is impossible also to compare meaningfully the small trailed combine harvester of 1934 with the monster of today that will gobble up some four acres of crop in an hour. The same could be said of the trailed two-furrow Ransome plough, and today's mounted seven-furrow reversible, or the Wilder Cutlift and forage harvesters now chewing up two acres of grass in an hour. On balance, change here must have been greatly to the advantage of the farmer, even if the strain on his cheque book has become almost unbearable.

Even in the livestock side of the industry, which has always been so much more labour intensive, things have changed drastically, though not perhaps at the same rate as with crop production. Modified buildings, livestock control systems, and materials handling equipment, now enable far larger groups of animals to be cared for swiftly by one man. This is exemplified by the extreme case perhaps of a man hand-milking some 16 cows sitting on a stool, compared with milking over 200 cows through a modern parlour. Even out in the field, the mini motorbike has come to the aid of the shepherd's legs, though we have not, one is glad to say, yet reached the stage of the electronic dog! That really would be a change for the worse.

It is a very different world for the stockman, and a better one, too, for the removal of many of his most backbreaking tasks is making life much easier and healthier for him. But he still has the satisfaction that comes from close contact with his stock – cows still have to be calved, ewes lambed, sick animals personally treated, and calves fed. Mechanisation has thankfully not completely removed the close contact of the stockman with his animals, nor will it ever do so. This bond between them is what makes livestock farming, in a way, more personally rewarding than the growing of crops. It is certainly very satisfying to see a clean, golden field of wheat, ready for harvest, with uniformly filled out ears of grain like an even carpet, or straw standing stiff and upright like guardsmen on parade. It is rewarding, too, to see a clean, uniform field of sugar beet covered with a thick, dark-green canopy of leaf, with not a weed in sight, and to know that it is manufacturing sugar at an astonishing rate in the bright sunshine. Rewarding certainly, but not to me giving quite so much satisfaction as the sight of an old sow grunting con-tentedly as she is suckled by a litter of 12 strong, pink piglets, or the crunching sound of 100 or so cows tucking into a nice new ley after turn-out from their morning milking.

Today's farming may not have quite the same old-world

aura given to it by writers in the past, but there is still much mental satisfaction to be gained from it. That is if there is time to seek it out in moments between filling up forms from the Ministry, or calls to the bank manager or the merchant to arrange credit until such time as the aforesaid Ministry sends you a cheque, weeks late because of a fault in a computer! It *is* a different world, naturally, but many of the fundamental things remain. The land must still be worked, and prepared for each new crop cycle, and much skill is required in spite of the advantages of power and the trains of implements at one's command. Today there is the greater challenge that our grandfathers did not have, of deciding when and in what quantities one should apply the wide range of chemical aids available to increment crop yield, without serious damage to the environment, and to ensure a financially optimum return – a challenge, indeed. And in the livestock world it is the same: the fundamentals of conception, birth, growth, production and the time of death remain. In spite of genetic manipulation, ovum transplantation, insemination, sire assessment, PIN and BLUP, we still have quotas, tuberculosis, sheep scab, scrapie, and BSE to prevent us from becoming too complacent. Such challenges still have to be met, and in the future there will undoubtedly be more to come in the field of ethics.

One is sometimes tempted to wonder whether all the scientific advances and increases in output may not have made the whole business of animal production a great deal more difficult to deal with, and sometimes with less benefit to the animal, than the simpler systems of 60 years back.

But then one recalls the endemic disease problems of those days, and, in addition, the hard unremitting repetitive work involved in livestock production. The mucking out of cattle yards and the manual handling of farmyard manure in the fields, the backbreaking humping of sacks of grain and feed, and the tedious hand milking of cows in the darkness of a winter's morning. No, one does not want to go back to that; so change, on balance, must be to our benefit – but not perhaps to quite such an extent as some of today's mechanisation enthusiasts would have us believe.

Most change is for the better, but we must never allow ourselves to be persuaded that *all* change is necessarily good. Change simply for change's sake should be resisted at all costs, and one could only wish that some of those in authority who control our lives would realise that fact, especially in education.

Fifty years ago an uncle of mine wrote his autobiography which he entitled, *It Passed too Quickly*. I would have liked to crib that title from him, for it expresses my own feelings about the past 60 years. They have simply whizzed by, and the alarming thing is that they continue to do so at an apparently ever increasing rate. That they have done so has been due, I am sure, firstly, to the wonderful luck that I have enjoyed in the job opportunities that were presented to me, and the variety of experiences that have sprung from them in so many different spheres. But, secondly, and more importantly still, it went so fast because of the enjoyment gained from working in agriculture, and from the help and assistance that I received from the scores of people whom I have met and worked with over the course of this fascinating journey. There can be no other industry in the world to equal it for kindliness and consideration for others. If you are listening up there, Dad, thank you once more for pointing me towards it those 60 years ago.

> *One generation passeth away, and another generation cometh, but the earth abideth forever.*
>
> Ecclesiastes 1:2

About the Author

After leaving Cambridge, Mike Soper worked at Reading, in the Sudan, and with the Surrey WAEC. From 1946–81, he lectured at Oxford, was Director of the University Farm, and became a Fellow of Christ Church. He was Honorary Secretary of the Oxford Conference from 1951–80, Chief Assessor for the National Certificate in Agriculture Examinations Board for 25 years, and Chairman of the City and Guilds Agricultural Committee from 1963–78, being elected an Honorary Fellow of the Institute in 1988. He chaired the Association of Agriculture Executive from 1984–90, and is currently Chairman of the Oxfordshire Agricultural Trust. He is also Honorary Secretary of the English Panel of the Council for Awards of Royal Agricultural Societies, and a member of the Advisory Body of the Cotswold Community.

FARMING PRESS BOOKS & VIDEOS

Below is a sample of the wide range of agricultural and veterinary books and videos we publish.
For more information or for a free illustrated catalogue of all our publications please contact:

**Farming Press Books & Videos
Miller Freeman Professional Ltd
Wharfedale Road, Ipswich IP1 4LG, United Kingdom
Telephone (01473) 241122   Fax (01473) 240501**

The Hired Lad   IAN THOMSON

A young man's first work on a Scottish farm when horses were yielding to tractor and bothy life was rough and ready.

The Spacious Days   MICHAEL TWIST

Growing up on a Buckinghamshire estate in the 1930s. Anecdotes about the farm staff, agricultural work, gamekeeping and the countryside.

Falling for It   GEOFF SURTEES

Geoff Surtees looks back to when, as a wide-eyed teenager, he took his first job as a forester on a Northumbrian estate, recalling the characters, wildlife and the time when the woods echoed to the sound of axes rather than chainsaws.

Early to Rise   HUGH BARRETT

An authentic and highly praised account of life as a farm pupil in the early 1930s.

Buttercup Jill   PEGGY GRAYSON

Amusing and entertaining memories of a pre-war rural childhood. Full of lively dogs, unpredictable horses and eccentric country characters.

British Farming: changing policies and production systems   ERIC CARTER & MALCOLM STANSFIELD

An introduction to the main types of farming enterprise, their current changes and their wider implications – ideal for schools.

Farming Press Books & Videos is a division of Miller Freeman Professional Ltd which provides a wide range of media services in agriculture and allied businesses. Among the magazines published by the group are *Arable Farming, Dairy Farmer, Farming News, Pig Farming* and *What's New in Farming*. For a specimen copy of any of these please contact the address above.